Pharmaceutical Analysis for Small Molecules

Pharmaceutical Analysis for Small Molecules

Edited by

Behnam Davani
Principal Scientific Liaison, General Chapters, Science Division,
U. S. Pharmacopeia

WILEY

This edition first published 2017

Registered Office
John Wiley & Sons, Inc., 111 River Street, Hoboken, NJ 07030, USA

Editorial Office
111 River Street, Hoboken, NJ 07030, USA

For details of our global editorial offices, customer services, and more information about Wiley products visit us at www.wiley.com.

Wiley also publishes its books in a variety of electronic formats and by print-on-demand. Some content that appears in standard print versions of this book may not be available in other formats.

Library of Congress Cataloging-in-Publication Data:
Names: Davani, Behnam, editor.
Title: Pharmaceutical analysis for small molecules / edited by Behnam Davani.
Description: First edition. | Hoboken, NJ : Wiley, 2017. | Includes bibliographical references and index. |
Identifiers: LCCN 2017029445 (print) | LCCN 2017030630 (ebook) | ISBN 9781119425038 (pdf) | ISBN 9781119425014 (epub) | ISBN 9781119121114 (cloth)
Subjects: | MESH: Pharmaceutical Preparations--analysis | Drug Design | Drug Approval | Chemistry, Pharmaceutical--methods | Small Molecule Libraries--pharmacology
Classification: LCC RS420 (ebook) | LCC RS420 (print) | NLM QV 25 | DDC 615.1/9--dc23
LC record available at https://lccn.loc.gov/2017029445

Cover image: (Pills) © Daryl Solomon/Gettyimages;
(Texture) © aleksandarvelasevic/Gettyimages
Cover design by Wiley

Set in 10/12pt, Warnock Pro by SPi Global, Chennai, India

Printed in the United States of America

10 9 8 7 6 5 4 3 2 1

This book is dedicated to my brother Behzad Davani.

Contents

About the Editor

Behnam Davani, PhD, has more than 25 years of experience in analytical chemistry, compendial science, QC/QA, and project management. He is currently Principal Scientific Liaison in the General Chapter Group, Science Division of the United States Pharmacopeia (USP). In this role, he coordinates the identification and scientific development of compendial (USP) courses for stakeholders worldwide. He is also active faculty for several compendial courses including method validation/verification/transfer, impurities in drug substances and products, compendial HPLC, residual solvents, stability studies for drug substances and products, spectroscopy, and others. He has taught these courses domestically and internationally, including in Canada, Europe, China, India, Russia, Korea, Latin America, Middle East, and North Africa. Prior to this position, he was Director of the Chemical Medicines Department. He provided scientific leadership and training to a team of international (India, China, and Brazil) and US-based scientific liaisons (PhD levels) responsible for the development and modernization of USP monographs and related general chapters for small-molecule drug substances and products.

Prior to this position, he was Senior Scientific Liaison in the Chemical Medicines Department (2003–2013). In this capacity, he worked with the USP Expert Committee, global pharmaceutical industry, and FDA for the development and revision of public standards for small-molecular-weight drug substances and drug products for human use. He was the Scientific Program Manager for the USP Industry Outreach Program (1999–2002). In this capacity, he managed the outreach program to enhance collaboration and communication with the major pharmaceutical companies and contract research organizations.

Prior to joining USP in 1999, he worked in various technical management positions in the industry for 12 years. He was Project Manager at Sigma-Aldrich for analytical method validation, stability studies, and method transfer in GMP Group/Pharmaceutical Division (1994–1999). He also managed the chromatography section in the Analytical Services Department, Research Division (1990–1994). Prior to that, he had management and research positions at HK

Environmental Services (1988–1990) and Midwest Research Institute (1987–1988). He also held the postdoctoral research fellow position for 1 year at the US Department of Energy (1986–1987).

Dr Davani has authored numerous technical publications and reports in the areas of compendial science, pharmaceutical analysis, and trace organic analysis with emphasis on chromatography and mass spectrometry techniques.

He holds a PhD in Analytical Chemistry from New Mexico State University and MS degrees in Physical and Analytical Chemistry from Cal Poly University, Pomona, and University of Iowa, Iowa City, respectively. He also has an MBA degree from Webster University in St. Louis, Missouri. He is a member of the American Chemical Society and the American Association of Pharmaceutical Scientists.

List of Contributors

Dr Motamed-Khorasani is a medical and scientific affairs expert and a senior scientist with a strong background in biomedical science and clinical trial/ research. She obtained her PhD and MS in molecular biology from University of Toronto and British Columbia, Canada, respectively, and did her Postdoctoral Fellowship at Microbix Biosystems Inc., Mississauga, Canada. She has a tenured and diverse range of experience in medical affairs, basic and industrial clinical research and development, clinical trials, medical and regulatory writing, and intellectual property. Dr Motamed-Khorasani has served as an independent consultant, director of medical affairs, senior medical sciences liaison, senior scientist, and senior medical analyst at Merck, Johnson & Johnson, United States Pharmacopeia (USP) Convention, Amgen, Baxter International, Covidien (eV3), Radiant Pharmaceuticals, AMDL Diagnostics, Microbix Biosystems, Neometrix Consulting, Samuel Lunenfeld Research Institute at Mount Sinai Hospital, Princess Margaret Hospital, and Vancouver General Hospital.

She has more than 20 years of experience and many national and international certificates in GLP, GMP, ICH-GCP, and FDA regulatory compliance for clinical trials and is a member of professional associations that include the Endocrine Society, American Association of Cancer Research (AACR), American Medical Writers Association (AMWA), Regulatory Affairs Professional Society (RAPS), American Society of Quality (ASQ), and Intellectual Property Institute of Canada (IPIC). Dr Motamed-Khorasani's research has focused on high-throughput approaches in the context of cancer informatics with a particular interest in the use of comparative analysis for the mining of integrated oncology datasets that include protein–protein interaction and gene expression profiling. She has published and presented more than 50 papers, abstracts, and articles in highly regarded scientific journals and high-profile conferences and scientific meetings.

Roger Dabbah, PhD, MBA, is the Principal Consultant at Tri-Intersect Solutions, which provides consulting, training in R&D and QA in the pharmaceutical, biotechnology, and medical device industries with specialization in microbiology testing, sterilization systems, and sterility assurance and management of laboratories. He also serves as an Adjunct Associate Professor at the University of Maryland University College, teaching courses in technology management including project management and risk management in projects, and at Johns Hopkins University, Whiting School of Engineering.

Prior to his current positions, he served as Director of the Complex Actives Division of the Drug Standards Department at the US Pharmacopeia. Before joining the USP, he was a Corporate Director of the Division of Microbiology, Sterilization, and Immunology for Baxter Healthcare, Manager of R&D Administration for the Nutritional Division of Abbott Laboratories, and Manager of the Biological and Information Sciences in technical services.

Dr Dabbah has extensively published papers and articles on microbiology, biotech, and management and collaborated in a number of technical books by contributing a number of chapters. He is also the author of two books: one on project management and the other on R&D management in the pharmaceutical industry. He is also on the Editorial Board of Pharmaceutical Technology and BioProcess International and on the Board of Directors of the PDA Foundation for Pharmaceutical Sciences.

Ernest Parente, PhD, is currently a Principal Scientist in the Chemistry, Manufacturing, and Control Regulatory Sciences group at Cardinal Health in Overland Park, Kansas. Formerly, he was a Sr Principal Analytical Chemist at Mallinckrodt Pharmaceuticals in St. Louis and the Head of Analytical Science and the Director of Quality Control at Sanofi-Aventis in Kansas City. In addition to his experience in Quality, Dr Parente has more than 18 years of experience in research and development and was the analytical chemistry team leader for the development of several currently marketed products. Before joining Sanofi-Aventis in 1989, he held positions in analytical and pharmaceutical R&D at Wyeth Laboratories, Warner-Lambert, and Hoffmann-La Roche. He has served at the USP for over 17 years and is currently a member of the USP Council of Experts. Dr Parente holds a PhD in analytical chemistry from the University of Delaware. He is an Adjunct Assistant Professor in the Graduate School of Pharmacy at the University of Missouri-Kansas City and is an active member of the ACS, Sigma Xi, AAAS, and AAPS. He is the author of scientific papers on chemical separations and protein analysis and has presented numerous US and International lectures on topics related to pharmaceutical analysis and the pharmaceutical industry.

Oscar Liu, PhD, is currently Director at Insys Therapeutics, Inc., leading pharmaceutical research and development including formulation. Prior to that, he has had several technical management responsibilities as Director and Senior Principal Scientist at Merck, Schering-Plough, Pfizer, and Par Pharmaceutical. In this role, he led Analytical Formulation/Product, Respiratory Product Development projects for more than 15 years. Dr Liu holds PhD degree in chemistry from Duke University. He has been board member of EAS since 2005 and was EAS president in 2014.

Dr Shaligram Rane has over 23 years in quality assurance/GMP/quality control and 2 years in academics with focus on streamlining and managing operations with proactive planning, changing existing or old concepts, and introducing new concepts for top-notch companies with consistent contribution to increased performance.

He completed his PhD in applied chemistry, MSc, and MEd. He has expertise in the quality and GMP department at various renowned organizations. Currently, he is heading the Quality (QC and QA) Department of Lupin Pharmaceuticals Ltd. (Biotech Division), Pune, India. Prior to Lupin, he associated with organizations such as Intas Pharma, Dishman Pharma, Cadila Pharma, Glenmark Pharma, Sun Pharma, Aarti Drugs and with Govt. Polytechnic College. His major areas of expertise are quality system, GMP activities, SAP-ERP system, designing of quality system according to regulatory guidelines. He has successfully handled more than 200 different types of inspection, for example, regulatory, customer, business partners, organization, and conducted more than 100 inspections at various pharma industries/laboratories. He has delivered talk on GMP topics at various workshops and conferences.

Dr Rustom Mody, Sr Vice President and Head of R&D (Biotechnology Division, Lupin Ltd.).

He has 19 years of experience in the Indian biopharmaceutical industry. He was the key person behind the development and commercialization of six biosimilars in the Indian market and two biosimilars in the US and EU markets. Dr Mody is currently developing four biosimilar products targeted for regulated markets such as the United States, Japan, Europe, and Australia. He is pioneer in the development and commercial-scale manufacturing of recombinant Hepatitis B vaccine.

Dr Mody has numerous patents filed, published, and approved and has 34 publications in peer-reviewed international journals. He was Ex-Chair of the Council of Experts for Biotherapeutics for United States Pharmacopoeia (Medicines Compendium) and Ex-Advisor to Indian Pharmacopoeia.

Preface

Pharmaceutical analysis is an important and integral part for the determination of quality including identity, purity, and strength of the drugs. In addition, related studies and programs are needed to assure the performance of the drug products. It requires analysts to acquire a solid understanding of analytical chemistry and also a thorough appreciation of pharmaceutical requirements to address these challenges.

The pharmaceutical industry is a major employer of science graduates, especially analytical chemistry majors. However, such students graduate with limited background in pharmaceutical analysis or related programs and are not prepared for employment in this industry. They find the transition from academics to this type of industry difficult due to lack of formal training in most of academic institutions. Therefore, this training/mentoring program is often performed by the pharmaceutical industry formally or informally to make these individuals productive employees as soon as possible. This type of training is also conducted as part of company's regulatory and quality programs. As a result, they found that there is a lack of introductory materials as they struggle to transition fast to new regulatory and more complex work environment.

This book is intended to be an introductory book for pharmaceutical scientists who are directly or indirectly involved with drug development process. It covers all major topics in pharmaceutical analysis, including related regulatory requirements. The book is useful for both new and experienced scientists, including analytical chemists, pharmaceutical scientists, quality control/quality assurance personnel, and pharmacists. It is also beneficial for students at undergraduate or graduate universities, schools of pharmacy in the United States and abroad for the pharmaceutical analysis course or online programs for regulatory science or quality control programs.

There are few other books/references in the area of pharmaceutical analysis. However, my goal is to deliver a concise and at the same time comprehensive book in this area. One way to achieve this is to focus only on the smaller-molecular-weight pharmaceuticals (drug substances and products).

The biological/biotechnological field and related analyses are beyond the scope of this introductory book and thus not covered.

The quality of pharmaceutical products must meet the required regulatory specifications, related guidelines, and good manufacturing and laboratory practices before being allowed to be marketed. Therefore, the book starts with the roles of FDA and ICH in setting such regulations and guidelines for drug approval process and submission (Chapter 1). Once specifications are approved, these become private standards enforced by FDA or other regulatory bodies. Chapter 2 extends this discussion to pharmacopeias and compendial approval process. This process leads to establishing public standards for pharmaceutical analysis by all stakeholders. Chapter 3 includes common methods for such analyses. The emphasis of this chapter is on more specific, stability-indicating, and instrumental techniques rather than classical nonspecific wet chemistry methods. Wet chemistry procedures are still used for routine analysis. However, the trend is toward automated instrumental tests for more sensitivity and specificity due to more stringer requirements for drug safety and toxicity concerns. This has also resulted in more efficiency and better characterization of the products, especially the determination of impurities at increasingly lower levels. The other focus of this chapter is on routine tests for the release and stability (QC lab) rather than more sophisticated instrumentation employed at the early stage in the research and development laboratory. The calculations associated with these analyses for both drug substances and products are included in Chapter 4.

The methods for pharmaceutical analysis have to be validated or verified if it is a compendial test. There is also a need to effectively transfer the noncompendial methods within the company or outsourced to qualified labs if needed. These topics are discussed in Chapter 5. Setting meaningful specifications and investigations in cases where these requirements are not met are discussed in Chapter 6. Due to the importance and more challenge to the analysis of impurities at trace or lower levels, a separate chapter is devoted to this topic (Chapter 7).

The remaining three chapters are related to GMP/GLP topics needed in a pharmaceutical regulatory environment. These include good documentation practices (Chapter 8), the management of analytical laboratories (Chapter 9), and analytical instrument qualifications (Chapter 10). These three chapters are placed at the end of the book. However, these are overarching chapters required during the entire life cycle of analytical procedures including development, validation, and performance verification. In addition, the list of abbreviations is included in both chapters and a separate appendix for the user's convenience.

I believe the order of the chapters flows logically for the pharmaceutical analysis. However, each chapter is written in such a way that is rather independent and can be referenced or studied separately. I hope that you will find reading of this book both useful and enjoyable. Comments and feedback to my email address bdavani1076@gmail.com are encouraged and appreciated.

February 2017
Behnam Davani

Acknowledgment

I would like to thank Professor Gary Eiceman, my PhD advisor at New Mexico State University, for the technical review of Chapters 3–5 and 7. His support and mentorship for many years during my career have been instrumental in my professional success.

I want to acknowledge the review of Chapter 3 and helpful suggestions by Dr Oscar Liu, who also contributed to a performance test section in the same chapter. I express my appreciation to my colleagues, Dr Horacio Pappa and Dr Andrzej Wilk, who supported this project as well as for many useful discussions we have had on various compendial topics during the last several years. I also want to express my gratitude to Dr V. Srini Srinivasan, former USP Chief Science Officer, for his support and encouragement as well as recommending contributors for the analytical instrument qualification chapter.

I am indebted to my late father, Hesam Davani, who planted the first seeds of my interest in learning, love of books, and critical thinking. I am grateful to my wife Ella Davani and my daughter Kimya Davani (Gheba) for their support and patience while making this long journey. This work could not have been completed without their constant encouragement.

1

Drug Approval Process and Regulatory Requirements

Roger Dabbah

1.1 Introduction

The role of the Food and Drugs Administration (FDA) in the review and approval of pharmaceutical products is divided into two broad categories. Each of the categories has its own set of regulations and issues, but regardless, they are designed to protect patients against harm and ensure the effectiveness of the medical products. These medical products include drugs from animals, plants, or human origin and products obtained via synthetic pathways, medical devices, and combination products

The two categories are as follows:

1) Analytics in the discovery process (R&D) of pharmaceutical products
2) Analytics in the compliance of products to their standards in the marketplace

Broadly speaking, the first category is a proactive approach while the second category is reactive. The first category ensures the safety and effectiveness of the products via the requirements for new drug applications (NDAs) and biologic license application (BLA) and occurs in the R&D phase of development of products, while the second category ensures that the manufacture of these products follows the NDA/BLA when they reach patients. The quality, safety, and effectiveness of pharmaceutical products are indicated via analysis of products that act as surrogates for these characteristics.

The nature of pharmaceutical products is their uniqueness that creates problems, issues, as well as challenges. A validated analytical procedure that works for one product might not provide for validation of the method for other prod ucts. The concept of validation must be applied in a flexible way to allow for changes due to the nature of a product, its chemical pathway, its origin and the nature of the APIs and inert ingredients (excipients) used for its manufacture. These issues will occur in both categories, and attempts to provide guidelines should include a more flexible approach that is not used presently.

Pharmaceutical Analysis for Small Molecules, First Edition. Edited by Behnam Davani.

The increase in regulatory requirements, often as a reaction to some perceived, potential, or real problems has increased the cost of development and compliance of pharmaceutical products. This is compounded by an adversary relationship among the regulatory agencies and pharmaceutical/biotech industry. In a perfect world, they should work in tandem in a win–win approach on scientific requirements and methodologies since they both have the same purpose, to ensure safety and effectiveness of pharmaceutical products.

However, before reviewing the role of FDA in the analytic areas, it would be of interest to briefly describe the FDA role by which a new drug entity is developed and approved.

In this chapter, we review in more detail the role of analytics required by FDA to approved products, to approve changes in products and to ensure through compliance that manufacture done according to NDAs will yield a quality product that is safe and effective. However, it is also important to discuss in some detail the good laboratory practices (GLPs) in 21CFR 58.

1.2 The Regulatory Process for New Drug Entity

A simplified schematic description of the overall FDA process [1] is shown in Figure 1.1.

1.2.1 Preclinical Studies

The organization will perform animal- or cell-based tests to determine if the drug is preliminarily safe and could become a candidate for human clinical trial. General guidance for these studies is provided by FDA, but must be adapted to the nature of the tested product. This is followed by one or more meetings with FDA, which reviews the data and, if necessary, requests additional data or clarifications. You can obtain guidances through the FDA website or through the Government Printing Office website.

1.2.2 Investigational New Drug Application (INDA)

The next step is to complete an INDA (21 CFR 312). Various "guidance for industry," some based on ICH, are available from the FDA website. Following review by FDA of the INDA and approval, clinical trials are conducted in Phase 1, Phase 2, and Phase 3

1.2.2.1 Phase 1 Clinical

Initial introduction of the investigational drug to 20–30 patients or normal volunteers to determine safety, pharmacologic actions, side effects associated with increased doses, and mechanism of action.

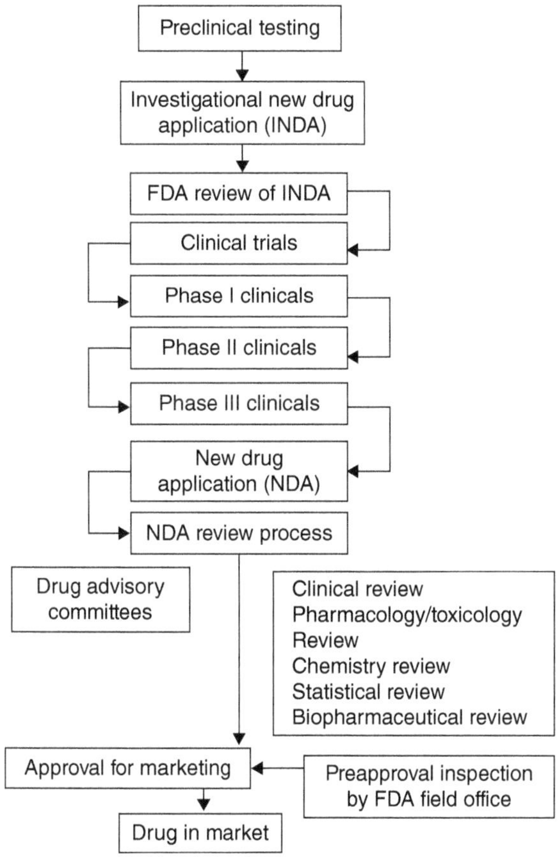

Figure 1.1 Schematic high-level representation of the overall FDA review process [1].

1.2.2.2 Phase 2 Clinical

Controlled clinical study to evaluate effectiveness and risks using hundreds of patients.

1.2.2.3 Phase 3 Clinical

Expanded trials to show effectiveness in several thousand patients.

1.2.3 New Drug Application (NDA)

1.2.3.1 NDA Review by FDA

For NDA with a high urgent priority, the review of the application will take about 6 months on average. For other NDAs, the target is to complete the review in 22 months.

1.2.3.2 NDA Review Process

See in Figure 1.2, a generalized NDA review process that was adapted by Dabbah [1] based on Mathieu [2].

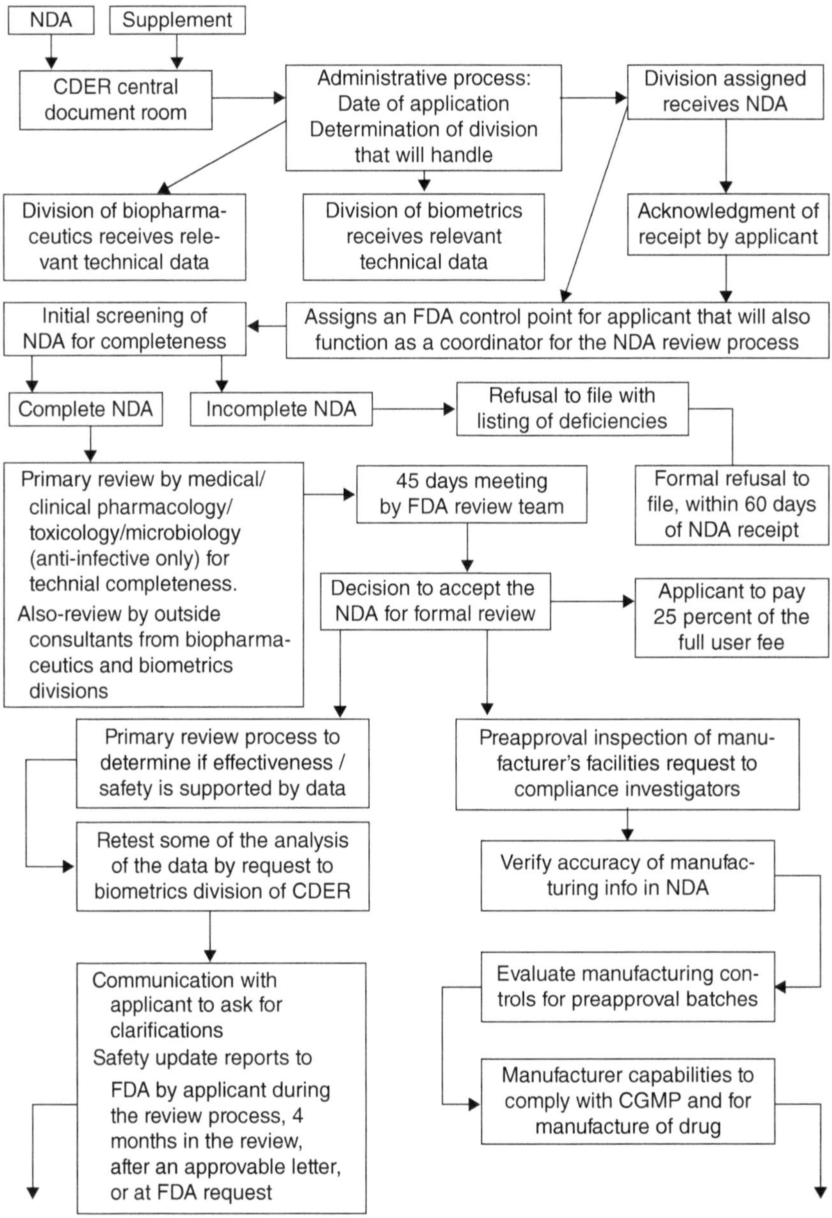

Figure 1.2 Generalized NDA review process from Dabbah [1], which was adapted from Mathieu [2].

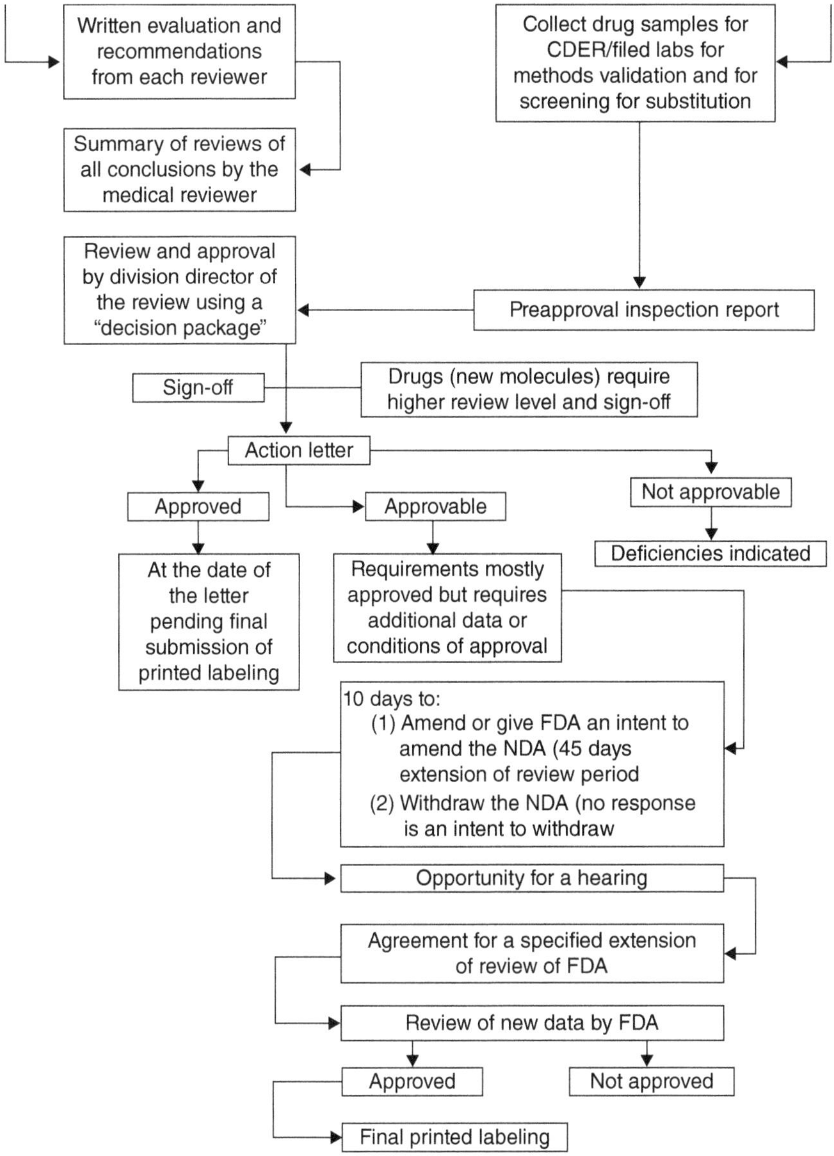

Figure 1.2 (*Continued*)

1.3 Good Laboratory Practice for Nonclinical Laboratory Studies

The intent of this section is not to reproduce the 21CFR-58 that one can obtain easily through the Internet, on the FDA website. The intent is to extract items that relate directly or indirectly to the analysis of pharmaceutical products, that would be applicable to products in development as well as to products that are on the marketplace. The scope of the regulation is large, but we will confine our discussion to human and animal drugs, medical devices for human use, and biological products. We will not discuss the animal facilities or the electronic products used [3].

The term of analytics applies to analysis of products using methods and procedures that have been validated for each of the products in question, and these tests are conducted according to protocols also called standard operating procedure (SOP) that would allow a consistent analysis of products. The results of analysis should ensure that the quality of the products fulfills the requirements of the NDAs for these products. If a test procedure has been validated, but the application of the test to the products is not done under a strict protocol, the credibility of the results will be in question, and the release of products to the marketplace will be harmful to patients and will also be illegal. The SOPs will include the environment of the laboratory where testing is being done. It goes without saying that an analysis must be performed by trained and skilled personnel under the supervision of the testing facility management or its delegate.

A requirement of GLPs is that there is a Quality Assurance Unit in the organization that will approve developed protocols designed to ensure the credibility of the results of analysis. Deviations in protocols must be approved by the Quality Assurance Unit before they are implemented.

Perhaps, one of the most important factors in assessing the credibility of analysis is the calibration of equipment for the purpose intended [4]. A credible analysis starts with the choice of a test article that should be representative of the tested system or the production batch. For example, in microbiological testing, microorganisms are not homogeneously distributed, thus representative sampling is a must. The use of control articles or reference standards is indicated in protocols to ensure that the tested article has the appropriate quality, strength, identity, purity, and composition to ensure the efficacy of the products.

The reporting of results of analysis must be based on the actual analysis of a product that is documented, archived, and retrievable.

1.4 Validation of Analytical Procedures: Methodology

Every analytical procedure must be validated. Guidance and recommendation are shown in Guidance for Industry: Q2B Validation of Analytical Procedures: Methodology, which was developed by the International Congress on Harmonization (ICH) and adopted by FDA in November 1996 [5]. Since these are guidelines, other approaches to validation may be acceptable.

The main objective of validation of an analytical procedure is to demonstrate that the test is suitable for its intended purpose. In general, one wants to determine the capability of the procedure in terms of specificity, linearity, range, accuracy, precision, detection, and quantitation limits. In Chapter 5, there is an extensive discussion of these characteristics applicable to most analytical procedures but might require some modifications due to the nature of the procedure and its applicability.

In the Guidelines for Industry on Validation of Analytical Procedures [6] (ICH-Q2A, which was also adopted by FDA (March 1995)), there is a general discussion of the seven characteristics shown earlier It also adds a section on the revalidation of the validation of analytical procedures. It should occur when there are changes in the synthesis of the drug substance, changes in the composition of the finished product, and changes in the analytical procedure.

The US Pharmacopeia information on the validation of analytical procedures should be consulted, inasmuch as that they are cited and applicable for products that are approved by FDA. These US Pharmacopeia (USP) chapters are <1223> Validation of Alternative Microbiological Methods [7]; <1225> Validation of Compendial Procedures [8]; <1227> Validation of Microbial Recovery from Pharmacopeial Articles [9] and <85> Bacterial Endotoxins Test [10].

1.5 FDA Role in the Discovery and Development of New Drug Entities

Each new drug, device, or biological is unique; thus, a single regulatory process that ensures safety and effectiveness is not desirable. Thus, the manufacturer of new entities must provide data on analytical procedures that include validation of analytical methods as well as adherence to GLPs as indicted earlier. If the FDA reviewer is not satisfied with the analytical data presented or the interpretations of these results, he/she might require additional data. It is a fact of practice that the manufacturer will not present all analytical data available, but only those that are required as a minimum. The approval process will go faster if manufacturers would provide to FDA all data that are available, even negative data. Small organizations as well as start-up organization that do not have too much experience dealing with FDA will tend to use the guidelines for industry to the verbatim, even when the nature of the new products is such that it does not require following these guidelines to the verbatim.

1.5.1 INDA Analytical Requirements

In this section, we look at the requirements for the development of analytical data. Before a drug entity is to be used for clinical trial, that is, administered to humans, the process includes an investigational new drug application (INDA). The INDA gives a general idea of pharmacological effectiveness and safety. The tests performed include screening via in vitro methodologies; pharmacodynamic testing via qualitative and quantitative pharmaceutical profile such as dose response, mechanisms of action, and interaction with other drugs; pharmacokinetics through bioavailability, accumulation, and clearance of the product and species to species differences. The assurance of safety is much more complicated and includes toxicological testing, via acute toxicity, subacute and chronic toxicity, carcinogenicity, reproductive toxicity, genotoxicity, and toxicokinetics testing. Each of the areas listed will yield credible and useful data if the procedures used are completely validated and follow the requirement of GLPs. Of more direct interest in this section in the INDA is the section on chemistry, manufacturing, and control (CMC). The chemical, physical, and biological characteristics of the drug are provided along with the validated analytical procedures that will be used to determine the identity, purity, potency, and, quality of the drug substance [11]. The information that is required depends on the phase of the investigation, risks, novelty of the drug, previous studies, route of administration, and the patient population targeted. At the INDA level, especially in Phase 1, there is a requirement for brief description of analytical procedures to be used. In the subsequent phases, there should be a list of tests performed, such as for the identification of impurities that should be qualified and quantified. If USP analytical procedures are used, they should be described in general terms. However, if non-USP analytical procedures are used, there is a need for a complete description including validation data [11]. The clinical investigation can start 30 days after the FDA receives the INDA application, unless FDA decides not to allow the start of the clinical phase. The reason for a hold on clinical investigation can be that FDA needs additional technical data, such as appropriate validation of the analytical procedures to be used or perhaps that the risk to patients is too high.

1.5.2 NDA Analytical Requirements

NDA requirements are covered in detail under 21 CFR Part 314. From an analytical point of view, there should be a description of analytical methods, their rationale for use, and appropriate statistical analysis. The CMC section includes references to the USP analytical methods as well as to non-USP analytical procedures with appropriate validation data. It is understood that both for INDA and NDA data presented to FDA have been obtained under GLP guidelines.

1.5.3 Biotechnology-Derived Products – Small Molecules

In 1999, ICH developed a Q6B guidance that was adopted by FDA as guidance for industry. It is titled Q6B Specifications: Test Procedures and Acceptance

Criteria for Biotechnological/Biological Products [12]. The objective was to provide guidance on general principles for the setting and justification of a uniform set of specifications for these products. Specifications, according to the guidance, are defined as a list of tests, reference to analytical procedures, and acceptance criteria. Conformance to specifications means that, when tested, using the analytical procedures indicated, these products will meet the acceptance criteria.

The analytical methods for biotechnology-derived products are very complex and mainly apply to large molecules. On the other hand, biotechnological processes can also lead to the development of small molecules, which will follow the requirements of drugs modified by the nature of the process and its process- or product-related impurities.

1.6 FDA Inspectors' Role in Analytics Relative to Products in the Marketplace

Drug manufacturing inspections are part of the overall involvement of FDA in ensuring the effectiveness and safety of products on the marketplace. The FDA has issued a number of guidance documents in its compliance program. We will review the general guidance for compliance, the guides for inspection of quality control laboratories, the biotechnology inspection guide, and the guide for inspection of microbiological quality control labs as they related to test procedures used, which constitute the overall area of analytics. The comprehensive regulatory coverage of all aspects of production and distribution of drugs that meet the requirements of the 501(a)(2)(B) becomes consistent across the pharmaceutical industry, thus reducing variations in compliance inspections.

1.6.1 FDA Compliance Program Guidance Manual (Implemented on 09/11/2015 with a Completion Date of 09/11/2016 – Program 7356.002)

The guidance manual [13] evaluates through manufacturer's inspections such as the collection and analysis of samples, the conditions and practices under which drugs and drug products are manufactured, packed, tested, and stored. Inspections are conducted every 2 years and zero in on compliance to current good manufacturing practices (cGMP)s. In this section, we deal with laboratory control systems. These include the availability of approved procedures and their documentations. The laboratory can have written approved procedures, but the role of the FDA inspector is to determine if the written procedures are used in the performance of analytical procedures. As the inspection proceeds, results might require a more in-depth investigation. For example, are the personnel qualified and trained to accomplish the various analytical procedures? Is the equipment available adequate and calibrated? If computerized or automated systems are used in connection with the analytical tests, are these systems validated? Are reference standards used in the analysis or are the standards used equivalent to the official reference standards? Are the required tests performed on the correct samples? Are all the records of the performed testing available? What is the out-of-specification (OOS) procedure and is it followed? Of major interest to FDA is the issue of the adequacy of samples used for analytical testing [14].

The FDA inspectors, often, are faced by the following issues in the laboratory control system that need to be remedied:

a) Pattern of failure to establish and follow a control system for implementing changes in laboratory operations
b) Pattern of failure to document investigation of discrepancies
c) Lack of validation of computerized and/or automated processes
d) Pattern of inadequate sampling practices
e) Lack of validated analytical methods
f) Pattern of failure to follow approved analytical procedures
g) Pattern of failure to follow an adequate OOS procedure
h) Pattern of failure to retain raw data.

1.6.2 Guide for Inspection of Microbiological Pharmaceutical Quality Control Laboratories

This is a specialized field of analysis that is often a reason for recall of products due to potential nonsterility or the presence of microorganisms that can be dangerous to the public at large and especially to immune-compromised patients. This guide addresses many of the issues associated with the chemical aspects of laboratory analysis of pharmaceuticals, but we would be remiss not to examine it in detail [15].

a) For pharmaceutical products that are nonsterile, the total microbial count would be an indicator of contamination if they exceed a certain limit. USP, in its monographs, addresses this issue by including microbial counts in the monographs. It also includes, when appropriate, the absence of certain specified microorganisms. These were determined by taking into consideration the use of the product, the nature of the product, the route of administration, and the potential hazard to the user. An FDA inspector should carefully review the microbiological testing of topical drug products, nasal solutions, and inhalation products, which appear to have a number of microbiological problems [16]. USP provides analytical methods for specified microorganisms [17], but it is the responsibility of the manufacturer to determine the native microbial population and, based on a risk/benefit analysis, to test for other microorganisms using a validated microbiological method. The advent of automated microbiological systems for count and speciation of microorganisms requires that the analytical method used is equivalent or better than the USP procedure. Faced by this issue, an FDA inspector will request comparative data between the USP method and the method used by the manufacturer's laboratory.

b) A number of products contain preservative, especially products used as multidoses. The procedure used should include neutralizing agents; otherwise, the results of microbiological testing will not be valid. In terms of validation, the USP procedures are validated, by definition. However, each drug substance or finished product contains inert ingredients that might interfere with the microbial tests. It will be necessary for the FDA inspector to request data on the validation of a microbiological analysis for a particular drug substance or finished product.

c) Media are necessary in most microbial tests, and the process used for preparation and use of media must follow an SOP. For example, sterilization of a medium must be validated. The environment of a microbiological testing laboratory is critical and needs to be monitored and controlled to ensure credibility of the microbiological testing results. Cross-contamination of samples must be avoided or at least minimized.

d) Sterility testing [18] has been an issue in recall of product for potential nonsterility. The test itself is subjected to contamination, unless precautions are taken to avoid contamination of samples. Robotic systems have been used, but might introduce a false sense of security. Similarly to any other system, the robotic instruments operation should be validated.

e) Procedures for microbial testing of product might originate in the USP, bacteriological analysis manual (BAM), or other microbiological references. The FDA inspector will evaluate the methodologies used and make sure that the laboratory has the equipment and instruments to conduct the test available and calibrated.

1.6.3 Biotechnology Inspection Guide

This is a specialized area that requires an understanding of the science and its application to the manufacture of biotechnology-derived products. The same basic regulations and requirements are applicable to these biotechnology-derived products if they are as small molecules. The FDA inspector should review the following areas to determine if deficiencies exist [19]:

a) The training of the laboratory personnel should be adequate for the performance of specific and complicated analytical procedures.

b) Equipment maintenance/calibration and monitoring should be documented, and a maintenance schedule should be available. All analytical methods should be validated with the equipment and reagents specified in the analytical procedures.

c) Reference standards or reference materials should be well characterized, properly stored, and utilized during testing.
d) Laboratory operating procedures should be available and followed.

1.7 Conclusions

- Analytical procedures are central to the assurance of the safety and effectiveness of drug substances, drug product, and biotechnology-derived products as well as biological product.
- Development of data must be done under GLPs to be incorporated in requests for marketing approval.
- Validation of analytical procedures ensures that the purposes of analytical procedures are fulfilled and is credible, thus protecting the patients and the consistency of manufacturing batches.
- Laboratories that perform analytical procedures must fulfill strict requirements in terms of their environments, the documentation of testing, the training of analysts and their skill levels, commensurate with the complexity of the analytical procedure used.
- The suitability of an analytical method for testing a given product must be established.
- Changes in analytical procedures should be justified and, when significant, must be approved by FDA under an amendment to the NDA. Revalidation of analytical methods should be done routinely at specified intervals or when changes in manufacturing process and/or ingredients are introduced.

The publication of guidance by FDA to its inspectors in evaluating the analytical procedures and for conformance to the approved marketing orders gives to the manufacturers a heads-up on what to expect during FDA inspections. Actually, these manuals should be used by manufacturers in preplanning for actual FDA inspections.

References

1 Dabbah, R., Total R & D Management – Strategies and Tactics for 21st Century Healthcare Manufacturers. Interpharm Press, Inc. Buffalo Grove, IL, 1999 , Chapter 13, pp. 472, 504–505
2 Mathieu, M., New drug Development: A Regulatory Overview, Revised third ed. Paraxel, Waltham, MA. 1994, p. 182
3 FDA, Good Laboratory Practice for Non-clinical Laboratories Studies, 2015, 21CFR 58 Chap. 1. http://www.accessdata.fda.gov/scripts/cdrh/cfdocs/cfcfr/CFRSearch.

4 Analytical Instrument Qualification <1058>, USP 39-NF 34. United States Pharmacopeial Convention, Rockville, MD.
5 FDA, Guidance for Industry-Q2B Validation of Analytical Procedures: Methodology, 1996. http://www.fda.gov/cder/guidance/index.htm.
6. FDA, Guidelines for Industry-Text on Validation of Analytical Procedures (ICH Q2A).
7 Validation of Alternative Microbiological Method <1223>, USP 39-NF 34. United States Pharmacopeial Convention, Rockville, MD.
8 Validation of Compendial Procedures <1225>, USP 39-NF 34. United States Pharmacopeial Convention, Rockville, MD.
9 Validation of Microbial recovery From Pharmacopeial Articles <1227>, USP 39-NF 34. United States Pharmacopeial Convention.
10 Bacterial Endotoxins Test <85>, USP 39-NF 34. United States Pharmacopeia Convention, Rockville, MD.
11 Hoiberg, C. P., Guidance on CMC for Phase 1 and Phases 2/3-Investigational New Drug Application, DIA/China Conference, Beijing, China, 2011
12 FDA, Guidance for Idustry-Q6B Specifications: Test Procedures and Acceptance Criteria for Biotechnological/Biological Products, 1999. http://www.fda.gov/cder/guidance/index.htm.
13 FDA, Compliance Program Guidance Manual-Drug Manufacturing Inspections (Program 7356.002), 2015.
14 FDA, Guide for Inspections of Pharmaceutical Quality Control Laboratories, 1993. http://www.fda.gov./ICECI/Inspections/InspectionGuides/ucm074918.htm.
15 FDA, Guide to Inspections of Microbiological Pharmaceutical Quality Control Laboratories, 1993. http://www.fda.gov/ICECI/Inspections/InspectionGuides/ucm074914.htm.
16 Microbiological Examination of Non-sterile Products: Microbial Enumeration tests <61>, USP 39-NF 34. United States Pharmacopeial Convention, Rockville, MD.
17 Microbiological Examination of Non-sterile Products: Tests for Specified Microorganisms <62>, USP 39-NF 34. United States Pharmacopeial Convention, Rockville, MD.
18 Sterility tests <71>, USP 39-NF 34. United States Pharmacopeial Convention, Rockville, MD.
19 FDA, Biotechnology Inspection Guide, 1991, http://www.fda.gov/ICECI/Inspections/InspectionGuides/ucm074181.htm.

2

Pharmacopeias and Compendial Approval Process

Roger Dabbah

2.1 Introduction

The roles and responsibilities of US Pharmacopeia (USP) are unique in the pharmacopeias of the world. In order to understand these roles and responsibilities, it is necessary to trace the development of USP since its inception in 1820. The mission of USP has evolved during these years, but not without controversies. In the year 2020, USP will celebrate its 200 years of continuous presence.

This chapter discusses the USP history, its purpose, and its organizational structure. It dwells into the revision process, the various publications, and briefly its working relationship with Food and Drugs Administration (FDA). Details of that relationship will be covered in Chapter 3. We complete this chapter by an in-depth discussion of the pharmacopeial initiative in harmonization among the pharmacopoeias in Europe and Japan and the US Pharmacopeia. The discussion of harmonization will also include some detail on the relationship of that initiative and the International Congress on Harmonization (ICH) initiative.

2.2 USP History

The USP was created on January 1, 1820, by 11 physicians who met in the Senate Chamber of the US Capitol Building. Private citizens concerned about the quality of medicines created a compendium of drugs in use, streamlined their names, and provided formula for their preparations. For comparative purposes, FDA did not exist at that time. The compendium evolved as the US Pharmacopeia that documents standard for identity, strength, quality, and purity of pharmaceutical products, including biotechnology-derived products, biological, medical devices, gene and cell therapy products, vaccines and blood and blood products. It also provides reference standards to be used in tests and assays [1].

Pharmaceutical Analysis for Small Molecules, First Edition. Edited by Behnam Davani.

The publication schedule has also evolved since; from 1820 to 1942, it was published every 10 years; from 1943 to 2000, it was published every 5 years; and, starting in 2002, it was published annually.

In 1975, the USP bought the National Formulary from the American Pharmaceutical Association, which now includes excipient standards as well as reference standards for these excipients. The history cannot be complete unless we mention that USP-NF standards were recognized by the Federal Food and Drugs Act of 1906 and by the Federal Food and Drugs & Cosmetic Act of 1938.

The US Pharmacopeial Convention that publishes USP/NF is a private non-profit organization that establishes legally enforceable national standards for medical products, including dietary supplements that are quoted in the Dietary Supplement Health and Education Act of 1994 amendments to the Federal Food and Drugs & Cosmetic Act.

2.3 Evolution of the Mission of the USP

The mission of USP has evolved rather rapidly. For example, in 1991, the mission of the USP was as follows:

> To promote the public health by establishing and disseminating officially recognized standards of quality and authoritative information for the use of medicine and related articles by health care professionals, patients, and customers

In 2013, in USP36-NF 31 (2013), the mission was as follows:

> *To improve the health of people around the world, through public standards and related programs that help ensure the quality, safety, and benefit of medicine and food*

Note the drastic difference between the two mission statements. For example, the USP/NF is for people around the world as compared to people in the United States. Note also that food has been included in the mission statement. The expansion of the USP mission to the world has been controversial. In particular, development of large numbers of medicine compendia standards overlapping with USP-NF monographs caused frictions between USP and stakeholders in a particular industry. For this reason, USP recently discontinued this initiative, and the monographs developed and posted on the USP website under this program were removed. Instead, USP in close collaboration with FDA, industry, and other pharmacopeias including WHO plans the development of a limited number of non-US monographs (not overlapping with current USP-NF monographs) for essential medicines. The proposal for the first of these new non-US monographs (chlorhexidine gluconate topical gel) was recently published in PF for public comment.

2.4 The USP Organization

The USP organization is composed of six components, each playing a different role but, when integrated, becomes the USP organization and governance [2].

2.4.1 The USP Convention

It is composed of delegates nominated by their State Pharmaceutical Associations, Colleges of Pharmacy. These delegates are nominated by State Pharmaceutical Associations, by Colleges of Pharmacy, by Colleges of Medicine, and by State Medical Associations. To ensure global representation, delegates at large are nominated.

The USP Convention meets every 5 years, elects a Board of Trustees, elects the USP President and Treasurer, and elects a Council of Experts. The USP Convention gives directions to USP for the next 5 years, through resolutions that are voted by the convention.

2.4.2 The Board of Trustees

This board assumes the responsibility of the management of the business affairs, finances, and properties of USP. It translates the resolutions of the convention into strategic and operational directives to the Council of Experts.

2.4.3 The Council of Experts

The members are elected by the USP Convention for a 5-year term. In the 2010–2015 term, 23 members were elected as Chairs of Expert Committees. Each Chair then elects members of their Expert Committee through lists of volunteers nominated by others or self-nominated. These experts are generally physicians, scientists, pharmacists, and other health-care professionals from academia, industry, or governments. They do not function as representatives of their companies or governments, but are elected because of their expertise. They serve as unpaid volunteers, but must obtain permissions from their organization to serve on the Committees.

2.4.4 Expert Panels to the Council of Experts

If an expert committee needs advisory recommendations on a specific issue, the Chair of the Council of Experts may appoint an Expert Panel that will disband when their recommendations are made.

2.4.5 Stakeholder Forums and Project Teams

The Stakeholder Forums provide an exchange medium for information between USP and stakeholders. USP discusses its standard setting activities, and the stakeholders bring issues that they would like USP to consider. The forums may form project teams for selected topics.

Examples of North American Stakeholder Forums (United States and Canada) include forums on prescription/nonprescription; dietary supplements; food ingredients, and veterinary drugs. International Stakeholder Forums, include forums in India, Mexico, Brazil, and others. Because of the change in mission of USP to extend the standard work to all countries, Scientific Standard Symposiums are conducted in the United States, China, India, Latin America, and the Middle East/North Africa.

2.4.6 USP Staff

The USP staff facilitates the revision process and the administration of Expert Committees. There are about 700 personnel, mainly scientists and professionals as well as administrative personnel mainly in the USP headquarters in Rockville, MD. USP also has an account management office in Basel, Switzerland, and laboratory facilities in Hyderabad, India, in Shanghai, China, and in Sao Paulo, Brazil.

2.5 The USP-NF Revision Process

The USP/NF is in continuous revision through a transparent process. It is a process that includes public participation following publication of proposal for changes or for new products in the Pharmacopeial Forum (PF). The process of publication and public comments can be iterated as many times as necessary according to the Expert Committee requesting the revision. The revision process has been enhanced a few years ago by publication of the PF online that is accessible free to all interested parties. The relevant Expert Committee reviews all the comments and incorporates them in the proposal, if necessary, but gives the justification for accepting or rejecting the public comments. The final proposal is republished before it becomes official [3].

Under Standard Revisions, 90 days as a comment period are given after publication in PF. However, Accelerated Revisions process that include errata, Interim Revision Announcement (IRA), and Revision Bulletins become official quickly upon posting on USP website and then incorporated in USP-NF print and online as soon as possible.

2.6 Publications of USP

2.6.1 USP-NF

The current USP/NF is composed of four volumes [4]:

Volume 1: The major sections include General Notices, General Chapters, Reagents, and Reference Tables.
Volume 2: Includes General Notices and USP Monographs, A–I.
Volume 3: Includes General Notices and USP Monographs, J–Z.
Volume 4: Includes General Notices, Dietary Supplement Monographs, and NF Monographs.

2.6.2 Pharmacopeial Forum

It is the official USP publication for public notice and comments. In 2011, the PF went online, free of charge. It contains revision proposals and additions to USP/NF, including harmonization proposals. It also contains Stimuli Articles for which USP would like public comments.

2.6.3 Supplements

There are two supplements per year: one published in February and becomes official on August 1, and the other published in June and becomes official on December 1.

2.6.4 USP Spanish Edition

Starting in 2006, USP provided a Spanish edition of USP-NF.

2.6.5 USP Reference Standards

It is a catalog of available USP reference standards used in comparison testing of a USP monograph for compliance. It can be accessed at the USP website (www.usp.org). Each of the USP reference standards is highly characterized products, generally submitted by manufacturers, which undergo collaborative studies for their establishments.

2.6.6 Chromatographic Columns

Provides information needed to conduct chromatographic procedures indicated in USP-NF. It is maintained in PF accessible through the USP website.

2.6.7 USP Dictionary

Published annually, the USP Dictionary of USAN and International Drug Names contains the up-to-date UA Adopted Names of drugs, official USP/NF names, nonproprietary and brand and chemical names. Other information includes graphic formulas, molecular formulas, and weights.

2.6.8 USP Dietary Supplements Compendium

Combines in a single volume USP-NF standards for dietary supplements, standards and information from USP-NF and the Food Chemical Codex and is published every 2 years in the print format.

2.6.9 Food Chemical Codex

Compendium of internationally recognized monographs, standards, and tests for the purity and quality of food ingredients (preservatives, flavoring, coloring, and nutrients). It is published every 2 years with supplements every 6 months. It is available in print and electronic formats. Revisions are proposed in Food and Chemical Codex (FCC) Forum that can be accessed for free at forum.foodchemicalscodex.org.

2.6.10 USP Medicines Compendium

Online compendium included monographs, general chapters, and reference materials for suitable chemical and biological medicines and their ingredients approved by national regulatory authorities outside the United States. However, these monographs were discontinued in mid-2015 and no longer available on USP website. Instead, the focus was shifted to modernization of current USP-NF monographs and development of limited number of non-US monographs for essential medicines in close collaboration with FDA, industry, and other pharmacopeias.

2.7 Relationship between USP and FDA

USP standards are recognized by law and custom in the United States and in many counties of the world as indicated previously. The USP and NF are recognized as an official compendium along with the Official Homeopathic Pharmacopeia of the United States. The provisions of the FD&C Act on adulteration and/or misbranding as well as the Public Health Service Act for biologics ensure that medicines fulfill their identities, potencies, and safety and are used as a legal justification for noncompliance, and products can be recalled on these bases [5, 6].

USP has no authority for the enforcement of standards, but FDA is empowered to enforce them. Often, FDA indicated that USP standards are minimum standards of quality and thus might require additional testings that are based on the FDA approval of medicine via new drug applications (NDAs), abbreviated new drug application (ANDA), new animal drug application (NADA)/abbreviated new animal drug application (ANADA), or biologic license applications (BLAs).

USP standards should not be confused with quality control testing for batch release as they often are. An example is in the microbiological testing of non-sterile products, especially in the testing of specified microorganisms. The USP microorganisms that should be absent in these products vary with the nature of the products and their route of entry in the body (oral, patches, suppositories, etc.). It is not enough to test only for these microorganisms, but also for pathogens that might be present in specific products because of their origin or their manufacturing process. FDA will insist, as they should, that testing for the absence of other microorganisms be part of the QC testing. The aim of the quality standards in USP monographs is to achieve and ensure consistency of products regardless of manufacturers or of processes used. USP does not mandate testing; what it mandates is that each and every unit in a batch up to the end of its shelf-life, fulfills USP requirements, if tested. From a legal point of view, it is more effective for FDA to invoke the provisions of adulteration and/or mislabeling than to invoke current Good Manufacturing Practices (cGMPs) guidelines infractions.

2.8 USP and the Pharmacopoeias of Europe and Japan

USP is unique among the pharmacopeias, that is, it is the only nongovernment organization that is empowered by law to develop standards for identity, purity, safety, and potency of pharmaceutical products that are enforceable by regulatory agencies. The pharmacopoeias of Japan and Europe are both government entities and have less degree of freedom to speed up the harmonization process that we discuss in detail in the next section. In addition, the USP is self-financing through the sales of reference standards and its publications [7].

The USP deals with drug substances, drug products for both human and animal use, excipients, biological and biotechnology-derived products, vaccines, blood and blood products, gene and cell therapy, medical devices, dietary supplements, and compounded preparations. The pharmacopoeia of Europe mainly deals with drug substances and not with finished products.

2.8.1 The European Pharmacopoeia

It is an agency within the European Directorate for the Quality of Medicine, in the Council of Europe. Harmonization drafts have to be provided to its audiences in English and in French.

2.8.2 The Pharmacopeia of Japan

The pharmacopoeias of Europe and Japan are government entities that report to government agencies, thus, creating some bureaucratic obstacles to any harmonization initiative. In addition, each pharmacopeia, including USP, has different revision processes and approval systems that complicate the schedule for implementation of harmonized chapters and monographs.

2.9 Harmonization of Pharmacopeial Monographs and General Chapters

The term harmonization is generally defined as to accommodate, adapt, and agree (Funk & Wagnalls, Standard Dictionary of the English Language, NY, 1958). It is also defined as concord, reconciliation of contradictions, or agreement in fact (Library of Universal Knowledge, Franklin J. Meine, Chicago, 1955). On the other hand, it is also defined as to bring into agreement (Webster's New World Dictionary of the American Language-College Edition, The World Publishing Company, 1959) and as being in agreement in terms of action, sense, and feeling (Dictionary.com). It is interesting to note than nowhere is harmonization defined as identical. This is an important distinction that has unexpected consequences in the harmonization initiative of the US Pharmacopeia and the pharmacopoeia of Europe and Japan. As an aside, note that the spelling of pharmacopeia/pharmacopoeia is not harmonized.

In the pharmaceutical/biotechnology industry, the meaning of harmonization has gone into various iterations. It is defined as we go along to resolve issues that were not expected at the start of the initiative. Harmonized monographs or general chapters can be exactly the same in all the three pharmacopeias, they could be harmonized by various attributes but not for others, or they are not harmonized at all.

The vehicle that carried on the harmonization was the Pharmacopeial Discussion Group (PDG) composed of representatives of the European Directorate for the Quality of Medicine (Phar.Eu) in the Council of Europe, the US Pharmacopeial Convention, Inc., and the Japanese Pharmacopoeia, first in the Ministry of Health and Welfare, then in the Ministry of Health, Labor, and Welfare (MHLW).The first face-to-face meeting occurred in 1989. In 2001, the PDG granted observer status to the World Health Organization (WHO). Later, we will discuss the relationship between the ICH and the PDG harmonization initiative. Although the harmonization process is science-based, we cannot deny that one of the reasons for pharmacopeial harmonization is economics. Since the push for internationalization of pharmaceutical products market is very strong, it does not make economical sense to test the same products in the three areas of Europe, Japan, and the United States because their monographs are different. The concept was that testing once would be sufficient to ensure the safety of patients in the three different areas.

The definition of harmonization arrived to by the PDG was as follows (see Chapter <1196> Pharmacopeial Harmonization in USP 36-NF 31, Volume 31 2013 as well as subsequent USP-NF and Supplements):

> *A pharmacopeial general chapter or other pharmacopeial documents is harmonized when a pharmaceutical substance or product tested by the document harmonized procedure yields the same results and the same accept/reject decision is reached*

The aforementioned practical definition of harmonization, then, would require the pharmacopeias to indicate the nature of the harmonization level (interchangeability, harmonization by attributes, or nonharmonized) using an appropriate coding system. Since monographs and general chapters need to be updated because of changes in regulatory requirements or because of technological advances, it is important that a process to revise harmonized documents be developed and implemented. We will also distinguish between retrospective harmonization designed to update monographs or general chapters already in the pharmacopeias and prospective harmonization for new monographs or general chapters. For details on the pharmacopeial initiative, you can also consult the following website: http://www.usp.org/sites/default/files/usp_pdf/EN/USPNF/pdg_state_of_work_june_2013.pdf.

2.9.1 PDG Working Procedures

There has been a commitment from the pharmacopeias to respect the PDG working procedures, although, they are rather rigid and constitute at times obstacles to the completion of harmonization. However, in international dealings, it is best to provide a firm framework to ensure that each decision is scrutinized by experts from each pharmacopeia and by the public who is asked to comment on these proposals at most steps of the process. The PDG biannual face-to-face meetings were also supplemented by meetings of experts to deal with significant issues [8].

Stage 1: Identification

Since it is not possible to harmonize all of the contents of pharmacopeias, the PDG had to set priorities for monographs and general chapters to be harmonized. The priority was established based on the inputs of stakeholders of the pharmacopeias as well as industrial trade associations and regulatory agencies. PDG then nominated a coordinating pharmacopeia for a given monograph or general chapter, making sure that the distribution of tasks to member's pharmacopeias is balanced. Some pharmacopeias that had special interest in a given monograph or general chapter were also set as one of the criteria for distribution of work.

Stage 2: Investigation
For retrospective harmonization, the coordinating pharmacopeia would gather materials from each pharmacopeia, the types and grades of products being marketed in each region, and the current analytical tests and assays. The coordinating pharmacopeia then would develop a draft monograph or general chapter and a report giving the rationale for the proposal, including validation data, and limits proposed. There was a tendency for the coordinating pharmacopeia to favor its own version of the document to be harmonized. The proposed draft and the report are sent to the secretariat of each pharmacopeia

Stage 3: Proposal for Expert Committee Review
Appropriate expert committee members who received the proposal ad examine it through consultation by correspondence or through meetings if significant issues are flagged. The expert committee from each pharmacopeia sends comments and proposed changes to their own secretariat that transmits them to the coordinating pharmacopeia. Comments should be sent between 2 and 4 months of the receipt of the Stage 3 draft. The coordinating pharmacopeia reviews the comments, accepts or rejects them but also gives rationale for their decisions, and prepares a harmonized draft that is written in a global style. The draft and the report are then sent to each secretariat of the pharmacopeias. There appears to be no provisions for the other two pharmacopeias to comment on that draft, especially if they disagree with the rationale provided, before it becomes an official inquiry document.

Stage 4: Official Inquiry
The draft including the commentary and justifications are published in the forum of each pharmacopeia in a section called "International Harmonization." The style that is published will conform to each pharmacopeia's style for publication or can be published in a global style, if preferred. The three pharmacopeias have 4–6 months to gather comments from their readers. Each pharmacopeia analyzes the comments of their own readers and submits the consolidated comments to the coordinating pharmacopeia within 2 months of the comment period. The coordinating pharmacopeia reviews all the comments and develops a new draft accepting or rejecting the comments but also produces a justification for its decisions.

Stage 5: Consensus

Stage 5A: Provisional
Within 4 months of the receipt of that draft, the three pharmacopeias try to arrive at a final consensus document. If consensus is not reached, the coordinating pharmacopeia develop a 5A/2 draft taking into consideration the significant

and substantiated comments of the other two pharmacopeias. This revised draft is sent to the secretariat of the other two pharmacopeias. Within 2 months of receipt, comments of the other two pharmacopeias are sent to the coordinating pharmacopeia. The process cycles as many times as necessary until a consensus is reached. This is when harmonization by attributes is invoked. Specific symbols are used to indicate the features that are harmonized by three pharmacopeias, by two pharmacopeias, or by no pharmacopeia. If the revised draft is significantly different than the original 5A document, PDG might require republishing of the last draft in each of the Pharmacopeias' forums.

Stage 5B: Draft Sign-Off
The coordinating pharmacopeia sends the final revised 5A draft 4 weeks before a PDG meeting. The document is then signed at the PDG meeting but also includes nonharmonized features that are clearly marked.

Stage 6: Regional Adoption and Implementation
This stage document is independently implemented according to each pharmacopeia's procedure.

Stage 6A: Regional Adoption and Publication
Each pharmacopeia incorporates the "harmonized" document according to their own procedures.

Stage 6B: Implementation
The date of the implementation of a "harmonized" document varies within each of the three areas. It depends on the legal requirements within each area, the need for translation, and publication schedules. Harmonization is not achieved until the document becomes official in the three pharmacopeias.

Stage 6C: Indication of Harmonization
Each pharmacopeia indicates in a statement the status of the harmonized document, including the attributes that are harmonized. The nonharmonized features are indicated by a black diamond.

Stage 7: Inter-Regional Acceptance
Since it is important that the regulatory agencies in the three areas accept the harmonized documents, there is a dialog between PDG and ICH starting at Stage 6 of the pharmacopeial harmonization process. The specific ICH Committee involved is ICH-Q4B, and the coordinating pharmacopeia provides the "harmonized" documents to ICH Q4B. Publication of the harmonized document is then done by each regional authority.

Stage 8: Revision of Harmonized Documents
No unilateral revision of a harmonized document is allowed. Revision will occur under specific conditions such as public health and safety, insufficient

supply of critical products of appropriate quality availability, unavailability of specific analytical reagents or instruments, or advances in analytical methodologies that are more appropriate, accurate, or precise than the current ones.

PDG will decide the need for revision, appoint a coordinating pharmacopeia, and the same process from Stage 1 to Stage 7 is then initiated. In case of public health and safety issues, the PDG may authorize a shortened process, by eliminating stages.

The PDG working procedures are very bureaucratic, rigid, and cumbersome. However, it is very transparent and takes into consideration the cultural and publishing differences among the three areas.

2.9.2 Status of the Pharmacopeial Harmonization Initiative

Details of the status of pharmacopeial harmonization can be found in the current USP/NF, in the current European Pharmacopoeia, and in the current Japanese Pharmacopoeia. In the USP-NF, the status of harmonization is under Chapter <1196> Pharmacopeial Harmonization. In the other two pharmacopoeias, similar chapters are to be consulted. A summary of the status of harmonization presented for various types of products as of June 22, 2013, is given as follows. In addition, an up-to-date status report on the progress of harmonization can be found in http://www.usp.org/usp-nf/harmonization [9].

General method relevant to ICH Q6A

Chapter	Coordinating pharmacopeia	Stage
<711> Dissolution: Revision 3	USP	6
<701> Disintegration: Revision 1	USP	6
<905> Uniformity of Content/Mass: Rev.1	USP	6
<61> Microbiological Examination of Nonsterile Products: Microbial Enumeration Tests – Rev.1	EP	6
<62> Microbiological Examination of Nonsterile Products: Tests for Specified Microorganisms – Rev.1	EP	6
<1111> Microbiological Examination of Nonsterile Products: Acceptance Criteria for Pharmaceutical Preparations and Substances for Pharmaceutical Use	EP	6
<85> Bacterial Endotoxins Test – Rev.2	JP	6
<631> Color and Achromicity (Instrumental Method) – Rev.3	EP	3
< >Extractable Volume – Rev.1	EP	6
<788> Particulate Matter in Injections – Rev.1	EP	6
<281> Residue on Ignition – Rev. 2	JP	6
<71> Sterility Tests – Rev.1	EP	6

General chapters

Chapter	Coordinating pharmacopeia	Stage
<786> Particle Size Distribution Estimation by Analytical Sieving – Rev.1	USP	6
<616> Bulk Density and Tapped Density of Powders	EP	4 (Rev.3)
<645> Water Conductivity	USP	3 (Rev.1)
< > Gas Pycnometric Density of Solids	EP	6
< > Powder Flow	USP	6
< > Tablet Friability	USP	6
< > Inhalation	EP	4 (Rev.3)
<776> Optical Microscopy	USP	6
<811> Powder Fineness	USP	6
<846> Specific Surface Area	EP	6
< > Porosimetry by Mercury Intrusion	EP	6
< > Laser Diffraction Measurement of Particle Size	EP	6
<941> Characterization of Crystalline and Partially Crystalline Solids by X-Ray Diffraction	EP	6
< > Water–Solid Interaction	EP	6
<891> Thermal Analysis	EP	4 (Rev. 1)
< >Uniformity of Delivered Dose of Inhalations	EP	2
< > Microcalorimetry	EP	6
<699> Density of Solids	EP	6
<621> Chromatography	EP	3 (Rev.1)

Methods for biotechnology products

Chapter	Coordinating pharmacopeia	Stage
< >Amino Acid Determination	USP	6
<726> Capillary Electrophoresis	EP	6
< > Isoelectric Focusing	EP	6
< > Protein Determination	USP	3 (Rev.1)
< > Peptide Mapping	USP	3 (Rev.1)
< > Polyacrylamide Gel Electrophoresis	EP	4 (Rev.1)

Excipients

Monographs	Coordinating pharmacopeia	Status
Alcohol – Rev.2	EP	6
Dehydrated Alcohol – Rev.2	EP	6
Benzyl Alcohol – Rev.2	EP	6
Calcium Disodium Edetate – Rev.1	JP	1
Calcium Phosphate Dibasic – Rev.1	JP	6
Calcium Phosphate Dibasic Anhydrous – Rev.1	JP	6
Carmellose Calcium – Rev.1	USP	6
Carmellose Sodium	USP	3 (Rev.2)
Croscarmellose Sodium	USP	6
Microcrystalline Cellulose – Rev. 1	USP	6
Cellulose Powdered – Rev.1	USP	6
Cellulose Acetate – Rev.2	USP	6
Cellulose Acetate Phthalate – Rev.1	USP	6
Citric Acid, Anhydrous – Rev.2	EP	6
Citric Acid Monohydrate – Rev.2	EP	6
Crospovidone	EP	6
Ethylcellulose	EP	4 (Rev.2)
Hydroxyethylcellulose	EP	5A3
Hydroxypropylcellulose	USP	6
Hydroxyethylcellulose, Low Substituted	USP	3 (Rev.3)
Hypromellose – Rev.1	JP	6
Hypromellose Phthalate	USP	6
Lactose, Anhydrous – Rev.4	EP	6
Lactose Monohydrate	USP	3 (Rev.3)
Magnesium Stearate	USP	6
Methylcellulose – Rev.2	JP	6
Methyl Paraben – Rev.1	JP	6
Petrolatum	USP	4 (Rev.)
Petrolatum White	USP	4 (Rev.)
Polyethylene Glycol	USP	4 (Rev.3)
Polysorbate 80	EP	6
Povidone – Rev.1	JP	4
Saccharin – Rev.1	USP	5B

Monographs	Coordinating pharmacopeia	Status
Saccharin Sodium	USP	4
Saccharin Calcium	USP	6
Silicon Dioxide	JP	4 (Rev.)
Silicon Dioxide, Colloidal	JP	4 (Rev.)
Sodium Chloride – Rev.3	EP	5A
Sodium Starch Glycolate – Rev.3	USP	5B
Starch, Corn – Rev.3	JP	6
Starch, Potato – Rev.2	EP	6
Starch, Rice – Rev.1	EP	4
Starch, Wheat – Rev.2	EP	6
Stearic Acid	EP	6
Sucrose	EP	6
Talc – Rev.1	EP	6
Ethyl Paraben – Rev.1	JP	6
Propyl Paraben – Rev.1	JP	6
Butyl Paraben – Rev.1	EP	6
Glycerin	USP	3
Carmellose – Rev.1	JP	6
Calcium Carbonate	JP	3/4
Copovidone	JP	4
Gelatin, Gelling Type/Nongelling Type	EP	6
Glucose Monohydrate/Anhydrous	EP	4 (Rev.)
Mannitol	EP	6
Propylene Glycol	EP	3
Sodium Laurylsulfate	USP	4
Starch, Pregelatinized	JP	3 (Rev.1)
SWFI in Containers	USP	3 (Rev.3)
Lactose for Inhalation	USP	3
Isomalt	EP	6

2.9.3 Roles and Responsibilities of Major Stakeholders in Pharmacopeial Harmonization

The roles and responsibilities of the pharmacopeias are as follows:

1) Develop and use the harmonization process via scientific working group with representatives of the three pharmacopeias.

2) Provide an administrative framework via their individual secretariat.
3) Scan the environment for new methodologies and techniques applicable to the work of harmonization.
4) Keep the harmonization process transparent at all time.
5) Provide the stakeholder forums with discussions of proposals by publications of drafts and the rationale for accepting/rejecting comments of the stakeholders.
6) Each pharmacopeia will bear the expenses of their representatives to the PDG and to the expert working groups.
7) PDG meetings will be held in Europe, Japan, or the United States in an alternate mode.
8) Provide consultation with regulatory agencies either directly or through the ICH.
9) Provide consultations with trade and scientific organizations such as Parenteral Drug Association (PDA), Biotechnology Industry Organization (BIO), American National Standard Institute (ANSI), American Association for the Advancement of Medical Instrumentation (AAMI), American Society for Testing and Materials (ASTM), International Pharmaceutical Excipients Council (IPEC), and other interested parties in the United States as well as their national equivalent trade and scientific associations in Europe and in Japan.
10) Do not make revision of harmonized documents without involving PDG and the process. For public health or safety reason and with the authorization of PDG, the process can be modified to take care of these issues.

2.9.4 The Roles and Responsibilities of Industry in Pharmacopeial Harmonization

1) Develop lists of excipients, monographs, and general chapters that have a high priority for pharmacopeial harmonization.
2) Work closely with the pharmacopeias to develop quality standards for new products. These also include the development of reference standards to be used in analytical procedures. Participation in the establishment of reference standards in round-robin testing. For harmonization to be as effective as possible, the industry should consider using the same reference standard for a new product in all three regions. A precedent exists in the development of a reference standard for the Bacterial Endotoxins Test.
3) Advise the pharmacopeias of new analytical procedural developments that could replace the current procedures with increase in accuracy, increase in speed of analysis and, when automated, could reduce the cost of testing. Provide validation data for these new procedures.
4) Provide timely comments and suggestions to proposals for harmonized documents that have been published in the forums of each pharmacopeia.

This can be done individually by each company and/or by trade and scientific associations.

5) Participate in open conferences sponsored by the pharmacopeias to discuss specific harmonization issues of interest.
6) Discuss with the regulatory agencies, in each of the three regions, the need for harmonization not only of quality standards and methodologies but also of the registration and approval process of new drug entities or modification of older drugs. Harmonization of quality and analytical standards by the pharmacopeias is only the first step in facilitating global commerce. It is necessary, but not sufficient, to reduce the cost of testing for regulatory compliance.

2.9.5 The Roles and Responsibilities of the Regulatory Agencies in Pharmacopeial Harmonization

1) Present the pharmacopeias with their priority in the harmonization of quality standards.
2) Attend open conferences and present the regulatory perspective on issues under discussions.
3) Avoid "podium regulations" that can affect the industry in the development of new products or modification of old products. Off-hand comments by regulatory agencies' representatives can create havoc in the industry and might necessitate planning changes for new products.
4) Comment on the various "harmonized" drafts that are published by the pharmacopeias in their forums.
5) Coordinate with other regulatory agencies to ensure that harmonization will be beneficial by facilitating the approval of new entities with a global perspective, namely harmonized methodologies, limits, and procedures.

2.9.6 The Roles and Responsibilities of the International Conference on Harmonization (ICH) in Pharmacopeial Harmonization

1) ICH provides a forum for the regulatory agencies and industry associations to meet and discuss issues of harmonization of registration and approval of new drugs. Among the various ICH working groups is the one dedicated to quality of these products.
2) The intersection of pharmacopeial harmonization and ICH harmonization work occurs in the quality working groups. The pharmacopeias are observers to these working groups.
3) When a "harmonized" draft is completed in Stage 6 of the pharmacopeial harmonization process, it is transmitted by PDG to the ICH Q group of interest. This allows ICH to comment before the final implementation of the harmonized documents is completed. This is, from the perspective of industry and of the regulatory agencies, a final check and balance on the

acceptability of the harmonized document from a compliance/regulatory point of view.

2.9.7 Advantages of Pharmacopeial Harmonization

1) One set of quality criteria that apply in the three regions.
2) Even the playing field thus expands the manufacturers' global potential marketplace.
3) In general, it would reduce the cost of testing products for compliance with the pharmacopeias.
4) By expanding the global marketplace, one can obtain ingredients, APIs, reagents, and instrumentation at competitive prices since a large number of suppliers will be available.
5) Analytical tests done for compliance to quality standards are not duplicated, reducing the time manufactured products can be introduced in the marketplace.
6) Harmonized documents (monographs or general chapters) apply to products manufactured by companies other than the "big Pharma" international companies. These documents can be used for generic products, OTCs, and products from small companies.
7) Reduces, somewhat, the burden of the regulatory agencies if they have only one agency reviewing the validation data for analytical methods used, instead of having multiple agencies doing the reviews if the analytical methods are not harmonized.
8) Decisions relative to harmonization are science-based, most of the times.

2.9.8 Disadvantages of Pharmacopeial Harmonization

1) Harmonization is too restrictive and does not cover a number of items used in pharmaceutical products. These include packaging materials, containers requirements, and instructions for storage conditions including the various definitions such as the temperature of storage. Labeling of dosage forms is not part of the harmonization work.
2) Changes in methodologies or limits brought about by harmonization require that each monograph that contains these tests must be modified. The task in enormous, since, for example, in the Bacterial Endotoxins test, there are over 700 monographs that have that requirement for testing. Furthermore, changes in heavy metal tests or residue for ignition are included in over 2000 monographs. The pharmacopeial process for changes in monographs or general chapters is long and laborious with publication in the PF in USP or equivalent forums for the other pharmacopeias. Comments from stakeholders after publication of a draft must be examined, accepted, or rejected, which justification for either decision.

3) The process for harmonization with seven stages is cumbersome and too bureaucratic. Contrast this with ICH process that includes only five stages.
4) Face-to-face meetings of PDG or of various working committees is time-consuming and expensive. Perhaps, new technologies for communication could be used since software that would allow experts in various location to work on the same document simultaneously exist.
5) Harmonized documents are either harmonized or not harmonized. However, there is a fallback position in the "harmonization by attributes." When consensus cannot be established, then the pharmacopeias agree to disagree on some of the features of a monograph or a general chapter. This is shown in the "harmonized document" using an appropriate symbol.
6) There are no guaranties that a harmonized document will be accepted by ICH and by the regulatory agencies.
7) Regional political factors can trump the science-based approach of pharmacopeial harmonization. From time to time, the regulatory agencies on behalf of the regional industry will try to introduce a regional advantage or reduce the probability of harmful effect of the new quality standards on their manufacturers. Political pressure is not unknown in all the three regions, and intense lobbying can affect the outcome of harmonization.

2.10 Comparisons between the PDG Process and the ICH Process in Harmonization

1) The PDG process has seven stages, while the ICH process has five stages [10].
2) PDG includes three pharmacopeias with WHO as an observer, while ICH includes the regulatory agencies in the three regions and the major trade associations in each region. The pharmacopeias have the status of observer, especially in the quality working groups.
3) PDG deals with all products, old and new, that are in the marketplace, while ICH deals only with new products.
4) Areas of the ICH quality working group that are of interest to the Pharmacopeias are as follows:
   ICH Q1A Drug Substance and Product Stability
   ICHQ1B Photo Stability
   ICHQ1C New Dosage Form Stability
   ICHQ2A Analytical Validation Terminology
   ICHQ2B Analytical Validation Method
   ICHQ3B New Drug Substance Impurities
   ICHQ3C Residual Solvent Impurities
   ICHQ4B Obtaining Regulatory Acceptance of Pharmacopeial Harmonized Documents

ICHQ5A Viral Safety in Biotech Products
ICHQ5B Genetic Stability in Biotech Products
ICHQ5C Biotech Product Stability
ICHQ5D Cell Substrates in Biotech Products
ICHQ6A Specification Tests and Acceptance Criteria for New Substances and Products
ICHQ6B Specifications for Biotech Products

2.11 The Special Case of Pharmacopeial Harmonization of Excipients

1) Excipients are difficult to harmonize since these substances are locally produced and their profiles including impurities and microbiological flora are quite different depending on the region where they are produced. An example would be sugar since it can be produced from beets or from cane. The same can be said about rice, wheat, starch, and corn, depending on the regional conditions of growing these products.
2) Excipient manufacturers are small manufacturers that cannot afford the latest instrumentation and procedure that are in the harmonized texts, since one of the purposes of harmonization is to update the methods based on new technologies.
3) Manufacturers of excipients have used wet chemistry methods to assess the quality of excipients. They are reluctant to change the tests, even if they can afford the new automated instrumentation and have personnel with the appropriate skill sets to run the analyses.
4) Acceptable limits for analytical tests can vary depending on the conditions of production. Perhaps, it would be wise to first harmonize the analytical methods and then deal with limits on a substance-by-substance basis.
5) Natural products have different profiles than synthetic products, which might create issues with the regulators.

2.12 Retrospective versus Forward Pharmacopeial Harmonization

1) Retrospective harmonization is relatively different and more difficult than forward harmonization. Current monographs and general chapters are the products of history and carry heavy luggage that can affect their harmonization. Development of tests and limits and their rational have, generally, been lost in time. Some might be based on the science available at the time of their developments. The reason for the acceptance of harmonization by attributes for excipients does not solve the issue of harmonization but only mitigates it.

2) Forward harmonization, which is the harmonization of quality standards for new products, is less difficult than retrospective harmonization. New products often use novel and new technologies for their analysis. There are no historical precedents to rely upon, thus allowing state-of-the-art instrumentations and procedures that are more accurate and generally with automation that increases the precision of the results. This requires that regulatory agencies be updated on the new technologies to be able to evaluate the results. Since forward harmonization is for new products, there is an intersection of interest between the PDG and ICH in harmonization that improves the probability of real harmonization in the three regions.

2.13 Conclusions and Recommendations

1) Pharmacopeial harmonization appears to be successful, especially for new products, such as biotechnology-derived products.
2) Critical general chapters, especially in the areas of microbiological testing, have been harmonized after numerous discussions among the pharmacopeias. These tests are frequently used for excipients, drug substances as well as finished products; thus, the impact of harmonization is felt rather widely in the industry as well as in the regulatory agencies.
3) The process used for the revision of harmonized texts appears to adequately work in maintaining harmonization even after changes are introduced.
4) The ICH process complements the PDG harmonization initiative by ensuring that harmonized texts are acceptable to regulatory agencies and to the industry.
5) There are other methods of ensuring harmonization of pharmacopeial texts: for example, a mutual acceptance by regulatory agencies in one region to accept texts (monographs, general chapters) of another pharmacopeia, even though they are not harmonized. Another method is harmonization by attributes that are used for excipients. The last method is not harmonization per se, but if consensus cannot be reached for all the attributes, then the pharmacopeias agree to disagree on some of the attributes. Finally, another approach called "Concordance of Foreign Pharmaceutical Tests and Assay" could be operating. This is a unilateral decision by one pharmacopeia to accept the texts from another pharmacopeia. This is done to facilitate global commerce, without having to go through the arduous process of PDG and ICH.
6) The PDG process can be streamlined, and unnecessary bureaucratic stages deleted. Which stages are to be deleted is a decision to be made by PDG.

2.14 Final Thoughts

Before the start of the PDG harmonization process in 1989, I visited several European Pharmacopeias to gauge their interest in the harmonization initiative. The head of one national pharmacopeia, a distinguished scientist, thought that attempting harmonization of pharmacopeia was similar to the travails of Sisyphus, a king in Greek Mythology that was punished by Hades to roll up a huge stone on a hill to reach the top of the hill. Once this was accomplished, Hades made sure that the huge stone rolled back to the bottom of the hill. The next day, Sisyphus started rolling the stone uphill until he reached the top of the hill. Hades made sure that the stone rolled down the hill, and the next day, Sisyphus had to roll the stone uphill. The scheme was repeated days after days. In pharmacopeial harmonization, one had the feeling that it was a travail of Sisyphus. However, contrary to Sisyphus, harmonization did achieve success, but it took a long and arduous trek up the hill.

List of Abbreviations

AAMI	American Association for the Advancement of Medical Instrumentation
ANADA	abbreviated new animal drug application
ANDA	abbreviated new drug application
ANSI	American National Standard Institute
ASTM	American Society for Testing and Materials (now: ASTM International)
BIO	Biotechnology Industry Organization
BLA	biologic license application
EP	European Pharmacopoeia
FCC	Food and Chemical Codex
FDA	Food and Drug Administration
ICH	International Conference on Harmonization
ICH-Q	International Conference on Harmonization–Quality
IPEC	International Pharmaceutical Excipients Council
IRA	Interim Revision Announcement
JP	Japanese Pharmacopoeia
NADA	new animal drug application
NDA	new drug application
NF	National Formulary
OTC	over the counter
PDA	Parenteral Drug Association
PDG	Pharmacopeial Discussion Group
PF	Pharmacopeial Forum
USP	United States Pharmacopeia

References

1 USP History in Mission and Preface, USP 36-NF 31, p. V. United States Pharmacopeial Convention, Rockville MD.
2 USP Organization in Mission and Preface, USP 36-NF 31, p. VI. United States Pharmacopeial Convention, Rockville, MD.
3 USP Revision Process in Mission and Preface, USP 35-NF 31, p. VIII. United States Pharmacopeial Convention.
4 USP Publications in Mission and Preface, USP 36-NF 31, p. IX. United States Pharmacopeial Convention, Rockville, MD.
5 USP Legal Recognition in Mission and Preface, USP 36-NF 31, p. Viii. United States Pharmacopeial Convention, Rockville, MD.
6 FDA, Federal Food, Drug, and Cosmetic Act, 21 USC, Chapter V, Jan. 2001 (in 2.7).
7 Pharmaceutical Harmonization <1196> – General Information, USP 36-NF 31, p. 942. United States Pharmacopeial Convention.
8 Pharmacopeial Harmonization <1196> – PDG Working Procedure, USP 36 - NF31, p. 942. United States Pharmacopeial Convention.
9 Pharmacopeial Harmonization <1196> – Status of the Pharmacopeial Harmonization Initiative, p. 944, USP 36-NF31. United States Pharmacopeial Convention.
10 International Congress for Harmonization – The ICH Process. Retrieved from www.ich.org/about/process-of-harmonization.html on August 18, 2015.

3

Common Methods in Pharmaceutical Analysis

Behnam Davani

3.1 Scope

The identity and purity of the pharmaceutical products must conform to the approved regulatory specifications before the pharmaceutical industry can market these products in the United States [1]. In this chapter, the most common techniques used for characterization of products are presented. A concise description of each method is provided, and advantages and disadvantages are discussed. However, the reader is encouraged to consult the references in this chapter for more detailed information about each technique.

3.2 Analytical Methods

3.2.1 Separation Methods

3.2.1.1 High-Performance Liquid Chromatography

High-performance liquid chromatography (HPLC) is the most widely used instrumental technology (35%) in analytical laboratories of pharmaceutical industry [2–6]. HPLC is more suitable for analysis of polar, water-soluble with relatively larger molecular weight compounds. Most of the pharmaceutical products are polar in order to be suitable for the recipient (patient) with water-based human body. This explains the popularity and common use of HPLC with polar mobile phase (e.g., water/acetonitrile) for analysis of most pharmaceutical products [7]. In addition, the use of several types of detectors including UV, fluorescence, refractive index, electrochemical, laser light scattering, aerosol-based detector, conductivity, and mass spectrometer has extended the applicability of HPLC for characterization of pharmaceutical products. Coupling of HPLC with other spectroscopic techniques such as FTIR, NMR has also provided excellent combination of separation and

Pharmaceutical Analysis for Small Molecules, First Edition. Edited by Behnam Davani.

structure elucidation of compounds of interest [3]. The four major modes of HPLC are as follows:

- Normal-phase chromatography (NPC)
- Reversed-phase chromatography (RPC)
- Ion-exchange chromatography (IEC)
- Size-exclusion chromatography (SEC)

NPC is also known as adsorption chromatography. In this mode, the stationary phase (typically silica and alumina) is more polar than the mobile phase. Therefore, the polar compounds are absorbed more strongly by the stationary phase and elute later compared to the nonpolar compounds.

RPC is more common than NPC. In this mode, the stationary phase (typically octadecylsilane or shorter alkyl chains) is less polar than the mobile phase. Therefore, the nonpolar compounds are adsorbed (retained) more by the stationary phase and elute later.

In IEC, the stationary phase contains ionic groups such as sulfonate (SO_3^-) or quaternary ammonium (NR_3^+), which are bonded to silica or resin. The mechanism of separation is based on exchange (displacement) of the counter ions (e.g., Na^+ or OH^-) with the ionic analyte (A^+ or A^-) in the mobile phase as follows:

$$\text{Cation exchanger: Resin} - SO_3^- \; Na^+ + A^+ \rightleftarrows \text{Resin} - SO_3^- \; A^+ + Na^+ \tag{3.1}$$

$$\text{Anion exchanger: Resin} - NR_3^+ \; OH^- + A^- \rightleftarrows \text{Resin} - NR_3^+ \; A^- + OH^- \tag{3.2}$$

SEC separates the compounds based on their molecular sizes. A porous material is used for the stationary phase. The larger analytes elute first because these are excluded from the pores. On the other hand, the smaller molecules diffuse into the pores and thus elute later.

Other chromatographic methods for specific applications or separation include the following [3, 5, 8]:

- Chiral chromatography: For the separation of enantiomers based on chiral stationary phase or chiral mobile phase.
- Ion-pair chromatography: For the separation of both ionic and neutral compounds by using ion-pairing reagents to make the samples suitable for common reverse-phase HPLC analysis.
- Affinity chromatography: Separation is based on a stationary phase containing a receptor specific for certain samples such as proteins and lipids.

The most recent trend in the development of HPLC instrumentation is ultra-high-performance liquid chromatography (UHPLC). This technique provides faster analysis and higher resolution, and it is finding increased application in the pharmaceutical industry.

3.2.1.2 Gas Chromatography

Gas chromatography (GC) is the second most wildly used instrument (15%) in the pharmaceutical industry [6]. This technique is sensitive, provides high resolution, and is excellent for quantitation [9]. It is especially suitable for quantitation of residual solvents, which must be controlled in both drug substances and drug products [10]. However, it is not suitable for nonvolatile or thermally labile samples. Some of the nonvolatile samples can be made volatile by derivatization reactions. This is not desirable because this process is time-consuming, and there is a need to extract the analyte from the interfering compounds in the matrix before analysis [11].

The most common detector for GC is flame ionization detector (FID). It is a universal and sensitive detector but lacks specificity. The thermal conductivity detector is also a universal detector but is less sensitive. Other detectors such as electron capture detector (ECD) and nitrogen phosphorous detector (NPD) are specific for halogenated compounds and those containing nitrogen and phosphorous, respectively. However, the most powerful and useful detector is mass spectrometer, which provides positive identification and excellent quantitation [9, 12].

3.2.1.3 Thin-Layer Chromatography

Thin-layer chromatography (TLC) is a simple and inexpensive method. It is not considered as sensitive as HPLC or other instrumental chromatographic methods since the detection is commonly performed by visual comparison of the analyte of interest with a reference standard. However, TLC coupled with densitometer detector can provide more sensitive and reproducible results.

In addition, TLC provides other advantages over HPLC including simple sample preparation, high sample throughput, and specific detections including both UV and chemical detection methods [13]. Furthermore, enhancement in the preparation of TLC plates containing smaller particle size sorbent has led to the development of high-performance TLC (HPTLC) with better resolution and faster analysis [4].

3.2.1.4 Supercritical Fluid Chromatography

Supercritical fluid chromatography (SFC) is used with either packed or capillary columns and a pressurized supercritical fluid such as carbon dioxide as mobile phase [14, 15]. It is mostly suitable for the analysis of nonpolar compounds because the most commonly used mobile phase is carbon dioxide, which is nonpolar. However, the polarity of the mobile phase can be increased by adding polar organic modifiers such as methanol.

Because supercritical fluid has properties that are similar to those of both gas and liquid, GC and HPLC detection systems (e.g., FID and UV) can be used. Due to the lower viscosity of the mobile phase, flow rate at higher mobile phase can be employed, resulting in faster separation. Another advantage of SCF is its suitability for the analysis of thermally unstable compounds [14]. It also provides higher resolution and faster analysis compared to those achieved by normal-phase HPLC [16].

3.2.1.5 Capillary Electrophoresis

Capillary electrophoresis (CE) is a more recent technique than HPLC. The mechanism of separation is generally based on different charge-to-size ratios of the analytes. It provides several advantages over HPLC including higher resolution, speed, and use of smaller quantities of sample. This technique is considered complementary to HPLC and has been applied more increasingly to pharmaceutical analysis [17].

3.3 Spectroscopy Methods

3.3.1 Ultraviolet

Ultraviolet (UV) spectroscopy, with spectral range of about 200–400 nm, is used for both quantitation and identification. The quantitation is performed by measuring the UV absorbance at a specified wavelength as a function of the concentration for both the analyte and reference standard. The identification is typically evaluated by comparison of the UV spectra (absorbance vs wavelength) and comparison to the spectra for the corresponding reference standard. When UV is used as identification tool, it offers limited selectivity for identification because different compounds may have the same or similar spectra. For this reason, this technique is usually combined with other spectroscopic techniques such as IR for positive analyte confirmation.

3.3.2 Infrared

The infrared (IR) spectroscopy is one of the most widely used identification techniques for pharmaceutical products. The spectral measurement is at about 2.5–15 μm (spectral frequencies of 4000–650 cm^{-1}). This technique offers excellent selectivity for different organic functional groups and can also be used for quantitative analysis. The popularity and effective use of this technique have been advanced by ease of spectral comparisons to a large number of published IR spectral databases through electronic data processing. However, it is not considered a sensitive technique for quantitation of low levels of impurities. In addition, molecules with no change of dipole moments (e.g., diatomic molecules with the same atoms) are infrared inactive [18].

3.3.3 Raman Spectroscopy

When a material is irradiated by visible light of single wavelength, the scattered light of different wavelengths characteristic of the material is observed

(Raman effect). The Raman spectra depend on the change in polarizability of the molecule rather than the change in dipole moment as is the case for IR. Therefore, those compounds that are IR inactive may be identified using this technique. Thus, Raman spectra provide complementary information to IR spectra. Another advantage of Raman spectroscopy over IR is its ability to characterize compounds in the presence of water as there are little spectral interferences from the aqueous solvent [18]. The pharmaceutical and other practical applications of this technique have already been well documented [19, 20].

3.3.4 Nuclear Magnetic Resonance

Nuclear magnetic resonance (NMR) spectroscopy is a powerful tool for structure elucidation of organic compounds. It has been successfully used for the characterization of drug substances and drug products [21]. This technique has also been used for the identification of polymorphs in drug substances [22]. However, its use as a quantitative method especially for low levels of impurities is limited.

3.3.5 Mass Spectrometry

Mass spectrometry (MS) is a powerful analytical tool and has had significant impact on the analysis of pharmaceutical products [23, 24]. The principle of MS is based on production of gas-phase ions from the sample in an electric or magnetic field and separation of ions due to the mass-to-charge ratio. It is used for both quantitative and qualitative analysis and is excellent for structure elucidation. Due to wide application of MS during the last couple of decades, the databases of spectra for many compounds of interest are available and easily used for comparison. The coupling of MS detector with other separation techniques such as GC, HPLC, SFC, and CE has also increased the versatility and application of this technique.

3.4 Other Spectroscopy Methods

3.4.1 Atomic Absorption Spectroscopy and Inductively Coupled Plasma Spectroscopy

Atomic absorption (AA) is a sensitive and reproducible method and is extensively used for the determination of elements [25]. However, only single-element determination can be performed at each run because separate radiation source (hollow cathode tube) for each element is needed.

Inductively coupled plasma (ICP) is an emission spectroscopy technique and is widely used for trace analysis of metals and other elemental impurities in a variety of matrices including pharmaceutical products. One of the advantages of ICP over AA is the simultaneous analysis of multiple elements. However, it is a more expensive and complex technique than AA.

3.5 Wet Chemistry Methods

3.5.1 Titration

Titration is a well-established procedure and used extensively for a variety of analytes [26, 27]. The analysis is based on the quantitative reaction of the analyte with the reagent (titrant). The titrant is added typically with a burette to the known volume of the analyte. The equivalence point (neutralization point) is reached when the amount of titrant is chemically equivalent to the amount of analyte. This equivalence point is determined visually by observing the change in the color of the added indicator or instrumentally. The change in the indicator color represents the end point (completion) of titration for the analyte of interest.

The titration has been grouped into several categories [27]:

- Direct Titrations: This is typically suitable when there is a rapid and quantitative reaction between the analyte and the reagent.
- Residual (back) Titrations: This procedure is used when suitable titrant cannot be found for the analyte. In this case, an excess amount of titrant is used. The excess (residual) amount is then titrated with the second titrant.
- Complexometric Titrations: This is used when a stable complex is formed between the analyte and the reagent. One common example is the titration of metal ions with ethylenediamine tetraacetic acid (EDTA) reagent.
- Reduction–Oxidation (Redox) Titrations: This involves oxidation–reduction reactions. One common oxidizing reagent used is potassium permanganate, $KMnO_4$.

For water-insoluble organic compounds, nonaqueous titration can be the method of choice. Use of organic solvents extends the application of titration procedure to a larger number of analytes that are not suitable for aqueous titration.

Titration provides several advantages including simplicity, low cost, high precision, accuracy, and applicability to a variety of analytes. The major drawback is the lack of specificity, and thus, it is not suitable as stability-indicating assay. In this case, a complementary test such as HPLC is required for the separation of impurities and potential degradation products.

3.5.2 Loss on Drying (LOD)

This procedure determines the amount of volatile matter driven off under specified conditions [28]. This simple procedure involves heating the sample under specified conditions (temperature, pressure, time). The sample is weighed before and after the end of this period. The difference in sample weight is the volatile content. The volatile compounds include water and potential volatile organic compounds such as solvents that may have been used in the manufacturing of the drug substance or products.

The limit of water in pharmaceutical substances is normally controlled to minimize the product stability or microbial growth. The residual solvents in pharmaceutical products are also controlled due to the toxic nature of certain solvents. This determination is also needed to calculate as appropriate the results for assay on the dried basis for comparison to the corresponding acceptance criteria (see Eq. (3.1)).

3.5.3 Loss on Ignition (LOI)

The procedure determines the amount of test material volatilized under specified condition of temperature [29]. In this test, the sample is ignited for a period of time to reach a constant weight. The difference in weight before and after the period is the loss due to ignition. If needed, the assay result needs to be corrected for the percentage of LOI to be consistent with the corresponding acceptance criteria (Eq. (3.2)).

3.5.4 Residue on Ignition (ROI) or Sulfated Ash

In this test, the amount of residual material not volatilized from a sample is determined when it is ignited at a specified high temperature in the presence of sulfuric acid [30, 31]. The sample is ignited at a high temperature to a constant weight. The difference between the weight before and after this ignition period corresponds to the amount of residue or sulfated ash. This amount usually corresponds to the content of inorganic impurities in an organic substance.

3.5.5 Water Determination

The amount of water (moisture) is commonly determined and controlled in the pharmaceutical ingredients for product quality and stability [32, 33]. Sometimes, the substance may contain water of hydration. In this case, the test is performed to confirm the specific (theoretical value) of the water expected in the substance. The most common procedure for the determination of the amount of water is the Karl Fischer Titration. Both volumetric and coulometric Karl Fischer titrations are used with the latter procedure providing more sensitivity for the determination of trace amounts of water.

3.6 Performance Methods (Contributed by Oscar Liu)

3.6.1 Disintegration

Solid dosage forms must first dissolve before they can be absorbed [34]. For oral solid dosage forms, disintegration is the first step of the dissolution process. Disintegration testing is the measurement of the amount of time the dosage form takes to disintegrate completely. Disintegration testing was originally established as a minimal quality control testing. Disintegration is directly related to tablet compression. A long disintegration time normally indicates that the tablets are highly compressed. On the other hand, an unusually short disintegration time may indicate undercompression. Major pharmacopoeias established disintegration testing in the 1940s and 1950s (British Pharmacopoeia in 1945, USP in 1950). The current USP has a harmonized (harmonized among USP, JP, and Ph. Eur.) general chapter on disintegration [33]. Currently, disintegration is less commonly used. Instead, dissolution testing (see Section 3.6.2) is more commonly required. However, for highly soluble drugs, disintegration may serve as a surrogate for dissolution testing.

USP Chapter <701> describes the disintegration testing procedure. In brief, the testing article is placed inside a vertical tube. A 10-mesh stainless steel screen is fixed to the bottom of the tube to prevent the testing article from moving out of the tube. The tube and testing article are immerged in a medium at body fluid temperature (37 ± 2 °C). Typical immersing media are distilled water, 0.1 M HCl, phosphate buffer, simulated gastric fluid, and simulated intestinal fluid. The tube is moved 5–6 cm up and down at a prescribed rate (e.g., 30 times/min). A plastic disc may be used to ensure that the dosage form is immersed completely during the up-and-down movement. A disintegration test apparatus consists of six such tubes. Six articles are generally tested at once.

3.6.2 Dissolution

Dissolution is the process by which a solid solute enters into a solution [35–39]. For a drug to be absorbed and be efficacious, it must first dissolve in a solution. For this reason, dissolution testing is important from both a development and a quality control perspective. In dissolution testing, a drug product is immersed in a degassed medium of known volume at a constant temperature (typically 37.0 ± 0.5 °C). Agitation is applied to the system to create relative movement between the drug and the medium. Typical dissolution media are hydrochloric acid (0.01–0.1N, to mimic stomach fluid), buffer solution with and without surfactant (to enhance solubility), and simulated biological fluids. The solubility of the drug in the dissolution medium should be at least three times the concentration of the drug when the entire drug is fully dissolved in the medium. When this condition is met, it is called the "sink condition." The amounts of drug dissolved at specific time points are measured and compared against the specification limits.

During the development stage, it is possible and helpful to develop *in vitro in vivo* correlation (IVIVC). With IVIVC, it becomes possible to predict the *in vivo* performance from the *in vitro* dissolution data. Even without IVIVC, dissolution is still a useful tool for monitoring the formulation and process development. In most cases, it is a mandatory test required by compendia as a form of quality control testing. There are seven different setups specified in the *USP* [35], each of which is appropriate for testing certain dosage forms. For quality control testing, six units are typically tested, and the results are compared to the predefined limits. This is known as the stage 1 test. If stage 1 testing does not meet the specifications, stage 2 testing may be conducted. In stage 2 testing, additional six units are tested. If stages 1 and 2 testing do not meet the limits, a stage 3 test may be conducted in which additional 12 units are tested. For immediate-release dosage forms, typically only one time point is specified. For oral delayed-release dosage forms, two time points are specified. One is an acidic stage; the limit is typically not more than (NMT) 10% dissolved. The other time point is a buffer stage. For extended-release dosage forms, three or more time points may be specified.

Dissolution testing for inhaled drugs is currently not mandated, and it presents special challenges. For inhaled drugs to be absorbed through the lung, aerodynamic particle size should typically be NMT 5 μm. For the dissolution data to be meaningful, particle segregation by their aerodynamic particle sizes should take place before exposing the particles to a dissolution medium. Introducing segregated particles into a dissolution apparatus presents a great challenge. In addition, it is challenging to select a dissolution medium that is bio-relevant. Finally, the typical delivered dose of an inhaled drug is very small compared to that of oral dosages. Consequently, a more sensitive analytical technology is required to detect the dissolved drug at these low concentrations.

3.6.3 Uniformity of Dosage Units

Uniformity of dosage unit testing is performed to ensure that each dosage contains drug substance within a narrow range around the label claim [40, 41]. USP Chapter <905> states that the uniformity of dosage units can be demonstrated by either of the two methods: "Content Uniformity" or "Weight Variation." The USP provides criteria for when the *Weight Variation* method can be employed. For this method, readers are suggested to read the current USP [39] for more information. For content uniformity, 10 units are initially tested. If the results from this stage of testing do not meet the acceptance criteria, additional 20 units are tested, and the results are compared to the acceptance criteria.

For respiratory delivery drugs, most units deliver multiple doses. Delivered dose uniformity must be assessed across the units and within the units [41]. For example, for nasal sprays, metered-dose inhalers, or reservoir dry powder inhalers, the delivered dose will typically be assayed at the beginning, middle, and end of unit life to assess the intraunit uniformity of the delivered dose. Multiple units will be tested to assess the interunit uniformity. In practice, a unit is actuated, and a drug dose is collected into a container, for example, USP dosing tube [41]. The collected drug content is assayed. The doses between the beginning and middle and between the middle and end of unit life are actuated to a waste container. It is recommended that the actuations be automated to minimize the testing variability.

3.6.4 Aerodynamic Particle Size Distribution Analysis

Another important test for inhalation products is aerodynamic particle size distribution (APSD) analysis [41–45]. APSD is critical for pulmonary delivery drugs as it will impact the location of drug deposition, which is critical for locally acting drugs. Pharmaceutical aerosols released from medical inhalers have irregular shape and varying density. Location of particle deposition depends not only on particle size but also on its density and shape and other factors. Aerodynamic diameter is defined as the diameter of a sphere particle of unit density that has the same setting velocity as the particle of interest.

APSD of aerosols delivered by medical inhalers is commonly measured by cascade impaction. USP Chapter <601> Aerosols, Nasal Sprays, Metered-Dose Inhalers, and Dry Powder Inhalers [41] describes the details of various apparatus for aerodynamic particle size measurement including Andersen cascade impactor, Marple–Miller impactor, multistage liquid impinger, and Next-Generation Impactor. The Andersen cascade impactor (ACI) is widely used. Within an ACI, there are a series of stages, each with an array of holes. Air is pulled through the impactors, and the size of the holes becomes smaller and smaller, causing the airstream linear velocity to become larger and larger from the top stage to the bottom stage. Aerosol particles are fractionated based on their aerodynamic size. Large aerodynamic particles will deposit on the upper stage due to their inertial impact while smaller particles will stay entrained in the airstream and move to the next stage, which has smaller jets and a larger airstream linear velocity. At each of the remaining stages, the same fractionation process takes place until the smallest aerodynamic particles pass the last stage and reach the filter paper, which is used to retain the smallest particles.

There are three primary mechanisms for aerosol *in vivo* depositions: inertial impaction, gravitational sedimentation, and Brownian diffusion. Large particles are mostly deposited by inertial impaction in the upper airway (or in the case of *in vitro* testing, in the top stages of an impactor). When the aerodynamic diameter is smaller than 1 μm, Brownian diffusion is the dominating mechanism. Airstream linear flow rate affects the inertial impaction and gravitational sedimentation. Temperature and humidity impact the aerosol solvent/propellant evaporation and/or moisture absorption. Therefore, this aerodynamic particle fractionation process should be operated at specific volumetric flow rate, temperature, and humidity.

Under defined experimental conditions, the cutoff diameter of each stage of the impactor is calibrated. A cutoff diameter is the particle diameter for which 50% of particles with that diameter deposit and 50% move beyond the stage.

Sample deposition at each stage is extracted and quantified by HPLC or by other analytical means. From the mass and cutoff diameters of each stage, the APSD can be established.

From the APSD, other parameters can be derived. Mass median aerodynamic diameter (MMAD) is the D_{50} on the cumulative distribution curve (mass vs. aerodynamic diameter). Fine particle dose or mass (FPD or FPM) is often defined as the total mass of particles smaller than a specified diameter, typically 5 μm. Alternately, fine particle fraction (FPF) can be used. FPF is the mass percentage of FPM. It is hypothesized that there is a correlation between FPF, FPD, or FPF and the mass or fraction of aerosol deposited in the lung. Therefore, APSD testing is critical for pharmaceutical aerosol characterization in development and quality control.

ACI testing is labor-intensive and time-consuming. Next-Generation Impactor™ is more user-friendly and efficient. It is recognized by both USP and European Pharmacopeia [45–47].

3.7 Microbiological Methods (Contributed by Roger Dabbah)

3.7.1 Introduction

The determination of purity and safety of nonsterile and sterile pharmaceutical products is directly related to the provisions of the Federal Food, Drugs, and Cosmetics Act that defines adulteration of a product that does not conform to microbiological tests that are included in the US Pharmacopeia [48]. It also relates to the concept of mislabeling when a product labeled as sterile is shown not to be sterile or a nonsterile product labeled *USP* that does not conform to the microbial limits indicated in the US Pharmacopeia. Additional microbiological tests can be operational if they are included in approved new drug applications (NDAs) or biologics license applications (BLAs).

The nature of microbiological testing in pharmaceutical products is different from physicochemical tests in terms of their accuracy or precision. One expects accuracy of a physicochemical method such as an HPLC method to be less than 5% because we are dealing with molecules that are uniformly dispersed in products. However, microorganisms are not distributed uniformly in pharmaceutical products, and most microbiological tests are at best estimates of the number of microorganisms in a sample because the conditions for microbiological tests (sample size, media selections, time and temperature of incubation) are compromised. For each of the microbiological analytical methods discussed next, we will describe the purpose of the test, the approach used, the limitation of the test, and the interpretation of the results.

For compliance purposes, alternatives to the cited methods can be used, provided they can be shown to be equal or equivalent to the pharmacopeial test that is the "Official Standard" that is accepted by the Food Drug Administration (FDA) and some other regulatory agencies. These alternatives methods include the so-called rapid tests that have to be evaluated against the pharmacopeial tests.

3.7.2 Microbial Limit Tests

There are two types of microbial limit tests: one designed to determine the enumeration of microorganisms in a sample and another type to determine the presence or absence of specified microorganisms.

3.7.2.1 Microbial Limit Tests – Enumeration via a Plate Count

Purpose of the Test: To determine the quantitative enumeration of mesophilic bacteria that may grow under aerobic conditions [49–51].

Approach Used: The ability of the test to detect bacteria in the presence of the product must be established. This requires that any antimicrobial substance in a product be removed or inactivated or neutralized. The capabilities of the media used to support the growth of bacteria, if present in a product, must be determined. This is done by testing the growth promotion of media using test microorganisms such as *Staphylococcus aureus, Pseudomonas aeruginosa, Bacillus subtilis, Candida albicans, and A. brasiliensis.*

Two Media are Used: Soybean-casein digest agar, which will be incubated at 30–35 °C for 3–5 days for the determination of the total aerobic microbial count (TAMC); Sabouraud dextrose agar at 20–25 °C for 5–7 days for the determination of the total yeast and mold count (TYMC). The preparation of samples depends on the physical characteristics of the product.

Limitations of the Test: The TAMC and the TYMC are counts that are obtained under a standardized set of conditions of temperature and time. It will exclude microorganisms that cannot be detected under the specified conditions. By convention, the TAMC is the count of CFUs that will grow in the soybean-casein digest agar, even if they are yeasts and molds. By convention, the TYMC is the count of colony-forming units (CFUs) that will grow in the Sabouraud dextrose agar, even if they are bacteria. This double counting is one of the limitations of the test.

Interpretations of Results: Since the counts are accepted as microbiological quality of the raw material and final products and taking into consideration the inherent variability of microbial counts, a limit of 10 cfu/g will be extended by a factor of 2 to a count of 20 cfu/g and still be acceptable as being within the limit [52].

3.7.2.2 Membrane Filtration Method

Purpose of the Test: To estimate the microbial count of samples of raw materials and nonsterile finished products.

Approach Used: The preparation of samples is similar as earlier. The microorganisms are separated from the sample through filtration using 0.45 μm filters. The procedure is similar to the plate count in that suitability of the procedure for each product should be determined using the same approach discussed earlier. If there is an antimicrobial in the sample, you can add an inhibitor or you can repeatedly wash the filter with specified solutions. Incubation of the membrane on the agar is similar to the one for plate counts.

Limitations of the Test: The product must be able to be filtered. For example, insoluble material will not be a good candidate for the use of the membrane filtration method. Additives can be used to the product, provided that you have validated the procedure including the lack of significant toxicity against microorganisms that might be present.

The limitations that were indicated for plate counts will be operative for the membrane filtration procedure.

Interpretation of the Test: Similar to the interpretation for plate counts.

3.7.2.3 Most Probable Number (MPN) Procedure

Purpose of the Test: To estimate the microbial count of products that could not be estimated using the membrane filtration procedure or the plate count procedure.

Approach Used: Use at least three serial 10-fold dilutions of the samples prepared as earlier. For each dilution, add 1 g or 1 mL to 9–10 mL of soybean-casein broth and incubate at 30–35 °C for NMT 3 days.

Limitations of the Test: It is not very accurate for an estimate of TYMC and less precise or accurate than the membrane filtration and the plate count procedures. Surface active agents (i.e., Polysorbate 80) and inactivators of antimicrobial agents need to be added. For some products, the visual determination of growth could be difficult, and in these cases, you can subculture the content of the doubtful tubes in broth or on plate agar incubated for 1–2 days at the same temperature. The MPN test is generally used when the bioburden of the product is very low.

Interpretation of Results: The MPN per gram or milliliter is read in the appropriate MPN table.

3.7.3 Tests for Specified Microorganisms

Purpose of the Test: Determine the presence or absence of specified microorganisms in raw materials or nonsterile finished products. This is done in order to ensure compliance with the FDA regulations and USP requirements. The list of specified microorganisms includes bile-tolerant

Gram-negative bacteria, *Escherichia coli, Salmonella species, P. aeruginosa, S. aureus, Clostridia species, and C. albicans* [53–55].

Approach Used: Samples are prepared as indicated in the enumeration tests earlier. Antimicrobial activity is removed or neutralized, and the suitability of the media used must be validated using the following American Type Culture Collection (ATCC) microorganisms, *S. aureus, P. aeruginosa, E. coli, Salmonella enterica, Clostridium sporogenes, and C. albicans.*

The principle of the test is to allow the growth of the specified microorganism while at the same time inhibiting the growth of other microorganisms. These properties must be validated using the organisms cited earlier.

Limitations of the Test: As for the enumeration tests, the absence or presence of specified microorganisms is tested under the prescribed conditions of temperature, media used, and length of incubation. Some variants of the specified microorganisms might not be favored by the standardized conditions. Not all raw materials and nonsterile products need to be tested for the absence or presence of all the specified microorganisms. It depends on the route of administration of the products [52, 56, 57]. For example, oral nonsterile products, in general, will be tested for the absence of *E. coli*. For products for oromucosal use, gingival use, and cutaneous use, the absence of *S. aureus* and *P. aeruginosa* will be tested. For materials of bovine origin, the absence of *S. species* should be assessed.

Interpretation of Results: The absence of specified microorganisms per g or mL of product, provided that the procedures are validated, might require further identification of the isolated bacteria, using commercially available systems.

3.7.4 Sterility Test

Purpose of the Test: This test, when performed as indicated, fulfills the requirement that a product labeled sterile is indeed conforming to the label [58–60].

Approach Used: A specified number of samples are tested, based on batch size and depending on the type of products. The lack of growth indicates that the sample tested is sterile. Two media are used: fluid thioglycollate medium, which will favor the growth of anaerobes and some aerobic microorganisms when incubated at 30–35 °C for 14 days; soybean-casein digest medium, which will favor the growth of fungi and aerobic bacteria, when incubated for 14 days at 20–25 °C. Growth promotion testing of the media used needs to be validated using ATCC strains of *S. aureus, B. subtilis, P. aeruginosa, C. sporogenes, C. albicans, and Aspergillus brasiliensis.* The suitability of the procedure needs to be validated for each product. The procedure does not have to be validated every time a product is tested unless changes in the procedure or the products have been introduced.

There are two types of procedures that can be used. The recommended one is using a membrane filtration procedure. If the method cannot be validated, then you can use the direct inoculation procedure. Regardless of the procedure used, you have to neutralize or remove the antimicrobial elements present in the sample before incubating the sample. The method of preparation of samples varies depending on the characteristics and nature of the products to be tested.

Limitations of the Test: The test is a compliance test and will not indicate if the whole batch is sterile. Sterility is being validated when the sterilization cycle is developed. This is a common mistake made by manufacturers that based the sterility of their products on the sterility test. Limitations of media compositions and preparations of samples, as well as the temperature of incubation, are a compromise that would result in some "sterile" products to contain microorganisms that do not grow under the conditions of the test.

The samples are observed at regular intervals during the 14-day incubation. As soon as a positive growth occurs, investigation is started to ascribe the cause of the positive growth. Identification of the positive microorganisms is the first step to determine if it is a survivor of the sterilization process or a contamination during testing. Another limitation of the sterility test is that the sample tested is not a statistical sample; thus, you cannot make any inference on the sterility of the whole batch.

Interpretation of the Results: When the product tested interferes with the visual assertion of growth, you will need to subculture the positive samples to a fresh medium and incubate for not less than 4 days. The lack of evidence of growth indicates that the samples comply with the sterility test. If microbial growth is found, it does not pass the sterility test. There are no provisions for retesting the batch unless the first sterility test is invalidated, for cause. If invalidated, you start the test using the same number of samples as in the original testing.

3.8 Critical Factors Involved in Microbial Limit Tests and in Sterility Tests

The elements critical to the assessment of microbial quality of products, if they are controlled, will increase the credibility of microbial testing to enhance the decision-making process of the decision-makers. These critical elements ranged from aseptic techniques, control of media, control of the ATCC test microorganisms, control of the equipment used in the procedures, training of the laboratory staff, appropriate and traceable recording, and evaluation and interpretation of results. These are reinforced by Lucia Clontz in her book on Microbial Limit and Bioburden Tests [61], who listed the top 10 regulatory issues associated with quality control in microbiology laboratories. These were poor recordkeeping and documentation, no laboratory investigation for aberrant or failure data, isolates not identified, nonvalidated methods and processes, lack of compliance with procedures, lack of standard operating procedures (SOP), lack of equipment qualification, lack of calibration of instrumentation, inadequate microbial monitoring of the environments, and poor employee practices.

3.9 Harmonization of Pharmacopeial Procedures and Requirement

Microbiological limit tests and sterility tests as well as the bacterial endotoxins test were harmonized among the US Pharmacopeia, the Pharmacopoeia of Europe, and the Pharmacopoeia of Japan. The harmonization started in 1989 and was successfully completed in May 2009. Reasons for the long delays were scientific, political, and the historical development of microbiological procedures and requirements in each of the pharmacopeias [62].

The pharmacopeias worked closely with the International Conference for Harmonization. The harmonized chapters included <61> Microbial Examination of Nonsterile Products – Microbial Enumeration tests, <62> Microbial Examination of Nonsterile Products – Tests for Specified Microorganisms, <71> Sterility Tests, <85> Bacterial Endotoxins tests, and <1111> Microbial Examination of Nonsterile Products – Acceptance Criteria for Pharmaceutical Preparations and Substances.

3.10 Bacterial Endotoxins Test

Purpose of the Test: Detection or quantification of endotoxins from Gram-negative bacteria. These substances, if present in injections, for example, could produce a pyrogenic or febrile response in patients. This is why a sterile injection is labeled sterile and nonpyrogenic [63–66].

Approach Used: The test uses amebocyte lysate from the horseshoe crab (*Limulus polyphemus* or *Tachypleus tridentatus*). It also requires a USP endotoxin reference standard. The test can be qualitative or quantitative. The test has to be validated for each product since there are substances that could depress or enhance the reactions. There are three techniques that can be used to perform the test:

1) The gel–clot technique, based on gel formation between the horseshoe crab lysate and endotoxins. Differences in terms of the results among the techniques are minimal, but the gel–clot technique is considered prime in terms of regulatory compliance in case of different results.
2) The turbidimetric technique based on the development of turbidity due to the cleavage of an endogenous substrate and the measure of that turbidity.

3) The chromogenic technique based on the development of color after cleavage of a synthetic peptide–chromogen complex.

Limitations of the Test: The test is a compliance test and will not tell you that the whole batch is nonpyrogenic. The amebocyte lysate reacts with some beta-glucans in addition to endotoxins giving some amplified results that can appear noncompliant. All glassware used in the test need to be depyrogenated (250 °C/30 min in a hot-air oven). Materials and equipment that are not heat-stable present a problem in depyrogenation by heat.

Interpretation of Results: Qualitative and quantitative tests are calibrated against a USP reference standard for endotoxins in order to be able to provide results that can be expressed numerically. Each product that has a USP monograph requiring a bacterial endotoxins test includes a maximum permissible limit of endotoxins when the test is run according to the prescribed methods.

The endotoxin limit for parenteral drugs is calculated on the basis of dose administered using the formula K/M. K is a threshold pyrogenic dose of endotoxin per kg of body weight, and M is the maximum recommended bolus dose of product per kg of body weight. $K = 5$ USP-EU/kg for any route of administration except for intrathecal route of administration, where $K = 0.2$ USP-EU/kg body weight. For radiopharmaceutical products not administered intrathecally, the endotoxin limit is calculated using the formula $175/V$, where V is the maximum recommended dose in milliliter, while for radiopharmaceutical products injected intrathecally, use the formula $14/V$. For anticancer drugs administered on a square meter of body surface, use the formula K/M, where $K = 2.5$ USP-EU/kg, and M is the (maximum dose/m^2/h × 1.80 m^2)/70 kg.

3.11 Summary

Common physical, chemical, and microbiological methods used in pharmaceutical analysis were reviewed. These include both instrumental and noninstrumental (wet chemistry) methods for identification and quantitation of both purity (active ingredient) and impurities. The advantages and disadvantages of each method were discussed. Due to the diversity of compounds of interest (active material, excipient, related or unrelated impurities or components), it is not feasible to use one method for full characterization of a drug substance or product. This is one reason why multiple tests and procedures are used for regulatory submission or in compendial monographs to characterize the drug substance or product.

These specifications (tests, procedures, and acceptance criteria) should be considered the minimum requirement as multiple manufacturers may have different processes for the production of their products. Therefore, the impurity profile and other specifications might be different but acceptable for regulatory approval of equivalent products. The manufacturers would have a better knowledge of their processes and route of synthesis and thus are in a better position to develop methods to fully characterize their products.

In addition, performance tests such as dissolution and uniformity of dosage units were discussed. These tests are needed to further ensure product quality, consistency, and its intended performance. Only some of the more common techniques and performance tests were discussed in this chapter. The readers are encouraged to consult other sources such as the references included in this chapter for other relevant tests as well as more detailed information on techniques and theoretical principles.

References

1 FDA, Federal Food, Drug, and Cosmetic Act, 21 USC Chapter V, January 2, 2001.
2 W. M. Dong Modern HPLC for Practicing Scientists, John Wiley & Sons, 2006.
3 V. R. Meyer Practical High-Performance Liquid Chromatography, John Wiley & Sons. 2004.
4 S. Ahuja Chromatography and Separation Science, Academic Press, 2003.
5 U. D. Neue HPLC Columns, Theory, Technology, and Practice, John Wiley & Sons, 1997.
6 K.C. Associates, Analytical Instrument End User Study, Wilmington, DE: K.C. Associates, 1997.
7 J. M. Miller, and J. B. Crowther Analytical Chemistry in a GMP Environment, A Practical Guide, John Wiley & Sons, 2000.
8 L. R. Snyder, J. J. Kirkland, and J. L. Glajch Practical HPLC Development, John Wiley & Sons, 1997.
9 H. M. McNair and J. M. Miller Basic Gas Chromatography, Techniques in Analytical Chemistry, John Wiley & Sons, 1997.
10 ICH, Topic Q3C (R4) Impurities: Guideline for Residual Solvents, February 2009.
11 R.M. Smith Gas and Liquid Chromatography in Analytical Chemistry, John Wiley & Sons, 1988.
12 M. McMaster and C. McMaster, GC/MS, A Practical User's Guide, John Wiley & Sons, 1998.
13 S. Ahuja and K. M. Alsante, Handbook of Isolation and Characterization of Impurities in Pharmaceuticals, Academic Press, 2003.
14 S. Ahuja and S. Scypinski Handbook of Modern Pharmaceutical Analysis, Academic Press, 2001.
15 K. Anton, C. Berger, Supercritical Fluid Chromatography with Packed Columns: Techniques and Applications, New York: Marcel Dekker, 1997.
16 Burger T. A. and Wilson W. H. Anal Chem, 65: [1993]; 1451.
17 M.G. Khaledi, High Performance Capillary Electrophoresis, New York: John Wiley & Sons, 1998.

18 H. Günzler, H.-U. Gremlich, IR Spectroscopy, An introduction, John Wiley & Sons, 2002.
19 C. J. Frank Review of pharmaceutical applications of Raman spectroscopy, Chapter 6, in Analytical Applications of Raman Spectroscopy M. J. Pelletier, Ed., Oxford: Blackwell Science, 1999, 224–275.
20 I. R. Lewis and H. G. M. Edwards Handbook of Raman Spectroscopy, New York: Marcel Dekker, 2001.
21 I. Wawer, B. Diehl and U. Holzgrabe NMR Spectroscopy in Pharmaceutical Analysis, Elsevier, 2008.
22 Anwar J., Garling T., and Barnes P., J Pharm Sci 78: [1989]; 337.
23 E. de Hoffmann and V. Stroobant, Mass Spectrometry Principles and Applications, Second Edition, John Wiley & Sons, 2004.
24 L. Zhou, in S. Ahuja and M.W. Dong, eds., Handbook of Pharmaceutical Analysis by HPLC, Elsevier, Amsterdam, 2005.
25 L. Ebdon, An Introduction to Atomic Absorption Spectroscopy, Heyden, 1982.
26 R. A. Day, Jr. and A. L. Underwood, Quantitative Analysis, Prentice Hall, 1991
27 General chapter <541> Titrimetry, USP35-NF30, United States Pharmacopeial Convention, Inc., Rockville, MD.
28 General Chapter <731> Loss on Drying, USP35-NF30, United States Pharmacopeial Convention, Inc., Rockville, MD.
29 General Chapter <733> Loss on Ignition, USP35-NF30, United States Pharmacopeial Convention, Inc., Rockville, MD.
30 General Chapter <281> Residue on Ignition, USP35-NF30, United States Pharmacopeial Convention, Inc., Rockville, MD.
31 Chapter 2.4.14 Sulfated Ash, European Pharmacopoeia, 2013 Edition.
32 HYDRANAL® Manual, Riedel de Haen, Seelze, Netherlands, 1995.
33 General Chapter <921> Water Determination, USP35-NF30, United States Pharmacopeial Convention, Inc., Rockville, MD.
34 General Chapter <701> Disintegration, USP35-NF30, United States Pharmacopeial Convention, Inc., Rockville, MD.
35 General Chapter <711> Dissolution, USP35-NF30, United States Pharmacopeial Convention, Inc., Rockville, MD.
36 General Chapter <724> Drug Release, USP35-NF30, United States Pharmacopeial Convention, Inc., Rockville, MD.
37 J. Dressman and J. Krämer, Pharmaceutical Dissolution Testing, Taylor & Francis Group, 2005
38 Umesh V. Banakar, Pharmaceutical Dissolution Testing, Marcel Dekker, 1992
39 Handbook or Dissolution Testing, Royal Hanson and Vivian Gray, Dissolution Technologies, Inc., Third edition, 2004.
40 General Chapter <905> Uniformity of Dosage Units, USP35-NF30, United States Pharmacopeial Convention, Inc., Rockville, MD.

41 General Chapter <601> Aerosols, Nasal Sprays, Metered-Dose Inhalers, and Dry Powder Inhalers, USP35-NF30, United States Pharmacopeial Convention, Inc., Rockville, MD.

42 Marple V. R., Roberts D. L., Romay F. J., Miller N. C., Truman K. G., Van Oort M., Olsson B., Holroyd M. J., Mitchell J. P., and Hochrainer D., Next generation pharmaceutical impactor (a new impactor for pharmaceutical inhaler testing). Part I: design. J Aerosol Med 16(3): [September 2003]; 283–299.

43 Marple V. A., Olson B. A., Santhanakrishnan K., Mitchell J. P., Murray S. C., and Hudson-Curtis B. L.. Next generation pharmaceutical impactor (a new impactor for pharmaceutical inhaler testing). Part II: archival calibration. J Aerosol Med 16(3): [September 2003]; 301–324.

44 Marple V. A., Olson B. A., Santhanakrishnan K., Roberts D. L., Mitchell J. P., and Hudson-Curtis B. L., Next generation pharmaceutical impactor: a new impactor for pharmaceutical inhaler testing. Part III: extension of archival calibration to 15 L/min, J Aerosol Med 17(4): [Winter 2004]; 335–343.

45 Chapter 2.9.18. Preparations for inhalation: aerodynamic assessment of fine particles, European Pharmacopoeia, 2013 Edition.

46 Chapter 2.9.44. Preparations for nebulisation: Characterization, European Pharmacopoeia, 2013 Edition.

47 General Chapter <1601> Products for Nebulization—Characterization Tests, *USP* 35-*NF* 30, United States Pharmacopeial Convention, Inc., Rockville, MD.

48 FDA, Federal Food, Drug, and Cosmetics ACT, Jan 2, 2001.

49 General Chapter <61> *Microbiological Examination of Non-Sterile Products-Microbial Enumeration Tests,* USP35-NF30, United States Pharmacopeial Convention, Inc., Rockville, MD.

50 Chapter 2.6.12. Microbiological Examination of Non-sterile Products-Microbial Enumeration Tests, European Pharmacopoeia, 2013 Edition.

51 Chapter. 35.1. Microbial Limit Tests-Enumeration tests, The Japanese Pharmacopoeia, Sixteenth Edition, 2011.

52 General Chapter <1111>*Microbiological Examination of Non-Sterile Products-Acceptance Criteria for Pharmaceutical Preparations and Substances for Pharmaceutical Use*, USP 35-NF 30, United States Pharmacopeial Convention, Inc., Rockville, MD.

53 General Chapter <62> *Microbiological Enumeration of Non-Sterile Products-Tests for Specified Microorganisms,* USP 35- NF 30, United States Pharmacopeial Convention, Inc., Rockville, MD.

54 Chapter 2.6.13. Microbiological Examination of Nonsterile Products - Tests for Specified Microorganisms, European Pharmacopoeia, 2013 Edition.

55 Chapter 35.2. Microbial Limit tests-Specified Microorganisms, The Japanese Pharmacopoeia, Sixteenth Edition, 2011.

56 Chapter 5.1.4. Microbiological Quality of Pharmaceutical Preparations, European Pharmacopoeia, 2013 Edition.

57 General Information Chapter 7.0. Microbial Attributes, The Japanese Pharmacopoeia, Sixteenth Edition, 2011.
58 General Chapter <71> *Sterility Tests*, USP 35-NF 30, United States Pharmacopeial Convention, Inc., Rockville, MD.
59 Chapter 2.6.1. Sterility Tests, European Pharmacopoeia, 2013 Edition.
60 Chapter 54. Sterility tests, The Japanese Pharmacopoeia, Sixteenth Edition, 2011.
61 L. Clontz, Microbial Limit and Bioburden Tests - Validation Approaches and Global Requirements, Buffalo Grove, IL: Interpharm Press, Inc., 1998.
62 General chapter <1196> *Pharmacopeial Harmonization*, USP 35-NF 30, United States Pharmacopeial Convention, Inc., Rockville, MD.
63 General Chapter <85> *Bacterial Endotoxins Test*, USP 35-NF 30, United States Pharmacopeial Convention, Inc., Rockville, MD.
64 J. F. Cooper. Bacterial endotoxins test in microbiology in Pharmaceutical Manufacturing, R. Prince, Ed., Bethesda, MD and Surrey, UK: PDA and Davis Horwood International Publishing, Ltd, 2001, p. 537.
65 Chapter 4.01. Bacterial Endotoxins Test, The Japanese Pharmacopoeia, Sixteenth Edition, 2011.
66 Chapter 2.6.14. Bacterial Endotoxins Test, European Pharmacopoeia, 2013 Edition.

4

Common Calculations

Behnam Davani

4.1 Scope

Calculations are part of analytical procedures and experimental design to quantitate major and/or minor components and impurities in the pharmaceutical products for evaluation of the required acceptance criteria [1–3]. The purpose of this chapter is to provide some examples of common calculations/formulas used in routine quality control (QC) tests. These include formulas used for wet chemistry (e.g., titration, loss on drying, etc.) and chromatography and spectroscopy techniques. The calculations for more complex and specific analyses as well as for those used for biological matrices are beyond the scope of this chapter. Furthermore, the calculations for statistical and related measurements are not included in this chapter. This latter topic is covered comprehensively in other references [4–6].

4.2 Calculations (Quantitative Analysis)

4.2.1 Percent Loss on Drying (LOD)

$$\%\ \text{LOD} = \left[\left(W_u + W_c\right) - W_d\right] / W_u \times 100 \tag{4.1}$$

where

W_u = weight of the sample (g)
W_c = weight of the container (g)
W_d = weight of the dried sample and container (g)

Pharmaceutical Analysis for Small Molecules, First Edition. Edited by Behnam Davani.

4.2.2 Percent Loss on Ignition (LOI)

$$\%\ \text{LOI} = \left[\left(W_u + W_c\right) - W_i\right] / W_u \times 100 \tag{4.2}$$

where

W_u = weight of the sample (g)
W_c = weight of the crucible (g)
W_i = weight of the ignited sample and crucible (g)

4.2.3 Percent Residue on Ignition (ROI)

$$\%\ \text{ROI} = \left(W_i - W_c\right) / W_u \times 100 \tag{4.3}$$

where

W_i = weight of the ignited sample (residue) and crucible (g)
W_c = weight of the crucible (g)
W_u = weight of the sample before ignition (g)

4.2.4 Assay

4.2.4.1 Chromatography (HPLC, GC)

Example 4.1 Drug Substance

Acceptance criteria: 98.0–102.0% of drug substance

$$\%\ \text{Assay Result} = \left(r_u / r_s\right) \times \left(C_s / C_u\right) \times 100 \tag{4.4}$$

where

r_u = peak response of the drug substance from the sample solution
r_s = peak response of the drug substance reference standard (RS) from the standard solution
C_s = concentration of the drug substance RS in the standard solution (mg/mL)
C_u = concentration of the drug substance in the sample solution (mg/mL)

Example 4.2 Drug Substance

Acceptance criteria: 98.0–102.0% of drug substance calculated on the dried basis.

The acceptance criteria for assay in Example 4.1 is on "as is" basis. This is usually the case where the amounts of water, solvent, and inorganic residue are at low (trace) levels. However, the assay acceptance criteria are generally expressed as "on dried basis," "on anhydrous basis," "on solvent-free basis" or "on the ignited basis" to correct for high levels of such components. The following formula with correction factor is used to calculate the assay result in such cases:

$$\% \text{ Assay Result} = (r_u / r_s) \times (C_s / C_u) / (1 - A) \times 100 \qquad (4.5)$$

where

r_u = peak response of the drug substance from the sample solution
r_s = peak response of the drug substance RS from the standard solution
C_s = concentration of the drug substance RS in the standard solution (mg/mL)
C_u = concentration of the drug substance in the sample solution (mg/mL)
A = percent (in decimal) for LOD, water, solvent, or LOI as appropriate

Example 4.3 Drug Product

Acceptance criteria: 90.0–110.0% of the labeled amount of drug substance in the drug product.

The formula for the calculation of the assay result for the drug product is similar to the one used for the drug substance. However, the results are expressed as the percent of the labeled amount of the drug substance in the drug product as shown in Eq. (4.6):

$$\% \text{ Assay result} = (r_u / r_s) \times (C_s / C_u) \times 100 \qquad (4.6)$$

where

r_u = peak response of the drug substance from the sample solution
r_s = peak response of the drug substance RS from the standard solution
C_s = concentration of the drug substance RS in the standard solution (mg/mL)
C_u = concentration of the drug substance based on the labeled amount of the sample solution (mg/mL)

Example 4.4 Drug Product (Salt Form)

The drug product is sometimes a salt form (e.g., hydrochloric acid, sulfate, phosphate, etc.). However, the active pharmaceutical ingredient is the free base (without the counter ion). Therefore, a correction factor must be obtained as shown in Eq. (4.7) to calculate the assay result.

$$\% \text{ Assay Result} = (r_u / r_s) \times (C_s / C_u) \times (M_1 / M_2) \times 100 \qquad (4.7)$$

where

r_u = peak response of the drug substance from the sample solution
r_s = peak response of the drug substance RS from the standard solution
C_s = concentration of the drug substance (salt form) RS in the standard solution (mg/mL)
C_u = concentration of the drug substance (free base) in the sample solution (mg/mL)
M_1 = molecular weight of the drug substance (free base)
M_2 = molecular weight of the drug substance (salt form)

The following example illustrates the point:

Drug product (salt) = oseltamivir phosphate capsules
Oseltamivir phosphate: $C_{16}H_{28}N_2O_4.\ H_3PO_4$ (salt form)
Active ingredient: oseltamivir: $C_{16}H_{28}N_2O_4$ (free base)

Assume:
Acceptance criteria: 90.0–110.0% of the labeled amount of oseltamivir ($C_{16}H_{28}N_2O_4$) in oseltamivir phosphate capsules

$$\%\ \text{Assay Result} = (r_u / r_s) \times (C_s / C_u) \times (M_1 / M_2) \times 100 \quad (4.8)$$

where

r_u = peak response of oseltamivir phosphate from the sample solution
r_s = peak response of oseltamivir phosphate RS from the standard solution
C_s = concentration of oseltamivir phosphate RS in the standard solution (mg/mL)
C_u = concentration of oseltamivir based on the labeled claim in the sample solution (mg/mL)
M_1 = molecular weight of oseltamivir, 312.40
M_2 = molecular weight of oseltamivir phosphate, 410.40

4.2.4.2 Spectroscopy (UV, IR, etc.)

Example 4.5 Drug Substance

$$\%\ \text{Assay Result} = (A_U / A_S) \times (C_s / C_u) \times 100 \quad (4.9)$$

where

A_u = absorbance of the drug substance in the sample solution
A_s = absorbance of the drug substance RS in the standard solution
C_s = concentration of the drug substance RS in the Standard solution (mg/mL)
C_u = concentration of the drug substance in the sample solution (mg/mL)

4.2.4.3 Titration

4.2.4.3.1 Direct

$$\% \text{ Assay Result} = \{[(V_u - V_b) \times N \times F]/(N_t \times W)\} \times 100 \tag{4.10}$$

where

V_u = titrant volume consumed by the sample (mL)
V_b = titrant volume consumed by the blank (mL)
N = actual normality of the titrant (mEq/mL)
F = correction factor (mg sample/mL of titrant as the theoretical normality)
N_t = theoretical normality of the titrant
W = weight of the sample corrected for water, loss on drying, solvent, and so on (mg)

4.2.4.3.2 Residual or Back Titration

$$\% \text{ Assay Result} = \{[(V_b - V_u) \times N \times F]/(N_t \times W)\} \times 100 \tag{4.11}$$

where

V_b = titrant volume consumed by the blank (mL)
V_u = titrant volume consumed by the sample (mL)
N = actual normality of the Titrant (mEq/mL)
F = correction factor (mg sample/mL of titrant as the theoretical normality)
W = weight of the sample corrected for water, loss on drying, solvent, and so on (mg)

Example

Analysis: Transfer about 1 g of the sample to a container. Add about 40.0 mL of 1 N primary volumetric solution (VS). Add indicator solution, and titrate the excess with 1 N titrant VS. Perform a blank determination. Each milliliter of 1 N primary VS is equivalent to 34.98 mg of the sample.

Experiment: 998 mg of the sample was titrated with 0.9994 N titrant VS. 15.0 mL of 0.994 N primary VS was used in the procedure. The blank consumed 44.02 mL of 0.9994 N titrant VS.

$$\% \text{ Assay Result} = \{[(V_b - V_u) \times N \times F]/(N_t \times W)\} \times 100$$

where

V_b = 44.02 mL
V_u = 15.0 mL

N = 0.9994
F = 34.98
N_t = 1.000
W = 998 mg (corrected for water content, loss on drying, solvent, and so on, if appropriate)

$$\% \text{ Assay Result} = \{[(44.02 - 15.0) \times 0.9994 \times 34.98] / (1.000 \times 998)\} \times 100 = 101.6\%$$

4.2.5 Organic Impurities

4.2.5.1 Chromatography (HPLC, GC)

Example 4.6 Using Area Normalization

The following simple formula can be used to calculate the percent of impurities in the drug substances or drug products. In addition to its simplicity, this calculation does not require the use of reference standard, which may be expensive or not easily available.

$$\% \text{ Impurity Result} = (r_u / r_t) \times 100 \quad (4.12)$$

r_u = peak response of each impurity from the sample solution
r_t = peak response of total peaks from the sample solution

Example 4.7 Using Relative Response Factor

The formula used in Eq. (4.12) assumes same relative response factors for all the impurities. This assumption may be not accurate. The use of relative response factor is another approach for the calculation of impurities. This alternative is superior to the one used using Eq. (4.12) because a correction factor (relative response factor) is used to account for different responses for impurities. Relative response factor is defined as the ratio of the response of the impurity to that of an equal amount of the drug substance. Using this definition, Eq. (4.13) can be used for impurity calculation.

$$\% \text{ Impurity Result} = (r_u / r_s) \times (C_s / C_u)(1/F) \times 100 \quad (4.13)$$

where

r_u = peak response of any impurity from the sample solution
r_s = peak response of the drug substance RS from the standard solution
C_s = concentration of the drug substance RS in the standard solution (mg/mL)
C_u = concentration of the drug substance in the sample solution (mg/mL)
F = relative response factor for each impurity (F)

Example 4.8 Using Reference Standard

The use of reference standard for the calculation of impurities is the third alternative. This approach minimizes the variations (errors) obtained in the calculation of relative response factors due to differences in sample purity, matrix, and/or experimental conditions. In addition, different approaches are used for the calculation of relative response factor. These include one point (one concentration level for impurity and drug substance within the linear range), average of the results from multiple concentration levels, or ratio of the slope of linearity curve for impurity to that for the drug substance. These approaches may also lead to different results for the relative response factor. Therefore, it is recommended to use the physical reference standard to minimize these variations. Equation (4.14) is used for the calculation of impurities in the drug substance using the impurity reference standard.

$$\%\ \text{Impurity Result} = (r_u / r_s) \times (C_s / C_u) \times 100 \tag{4.14}$$

where

r_u = peak response of the impurity from the sample solution
r_s = peak response of the impurity RS from the standard solution
C_s = concentration of the impurity RS in the standard solution (mg/mL)
C_u = concentration of the drug substance in the sample solution (mg/mL)

The corresponding equation for the determination of the impurity in the drug product is as follows:

$$\%\ \text{Impurity Result} = (r_u / r_s) \times (C_s / C_u) \times 100 \tag{4.15}$$

where

r_u = peak response of the impurity from the sample solution
r_s = peak response of the impurity RS from the standard solution
C_s = concentration of the impurity RS in the standard solution (mg/mL)
C_u = concentration of the drug substance based on the labeled amount of the sample solution (mg/mL)

4.3 Calculations (System Suitability Parameters)

Calculations of some common system suitability chromatographic parameters are shown as follows [7–9]:

4.3.1 Resolution (R)

Resolution is a measure of the separation of two adjacent peaks and is calculated by the following equation (see Figure 4.1) [7]:

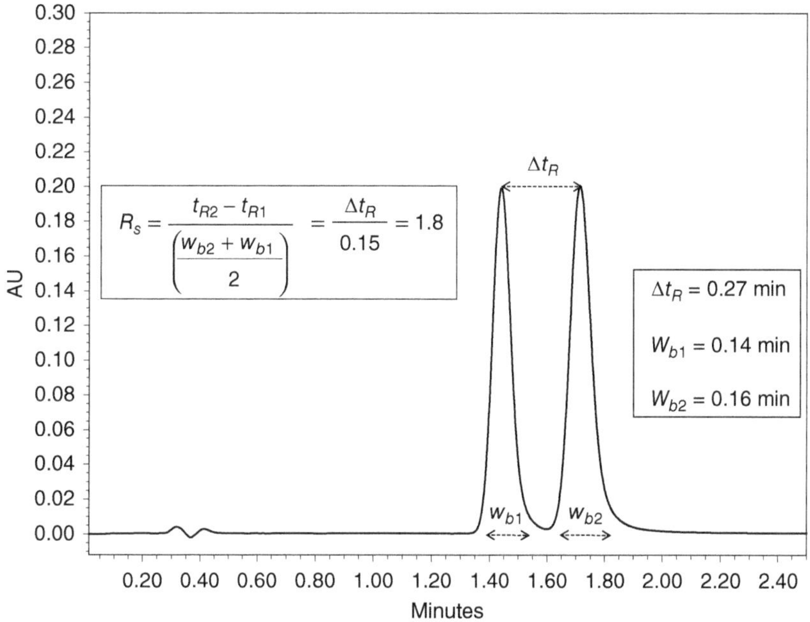

Figure 4.1 A chromatogram of two peaks with a resolution (R_s) of 1.8.

$$R = \frac{2(t_{R2} - t_{R1})}{(W_{b2} + W_{b1})} \tag{4.16}$$

where

t_{R2} and t_{R1} = retention time of the two adjacent peaks
W_{b2} and W_{b1} = peak widths measured at the baseline for the two corresponding peaks

4.3.2 Tailing Factor (*T*) or Asymmetry Factor (A_s)

These parameters are the measure of the peak symmetry and are defined by the following two equations (see Figure 4.2) [7]:

$$T = \frac{W_{0.05}}{2f} \tag{4.17}$$

where

$W_{0.05}$ = peak width at 5% of the peak height
f = see Figure 4.2

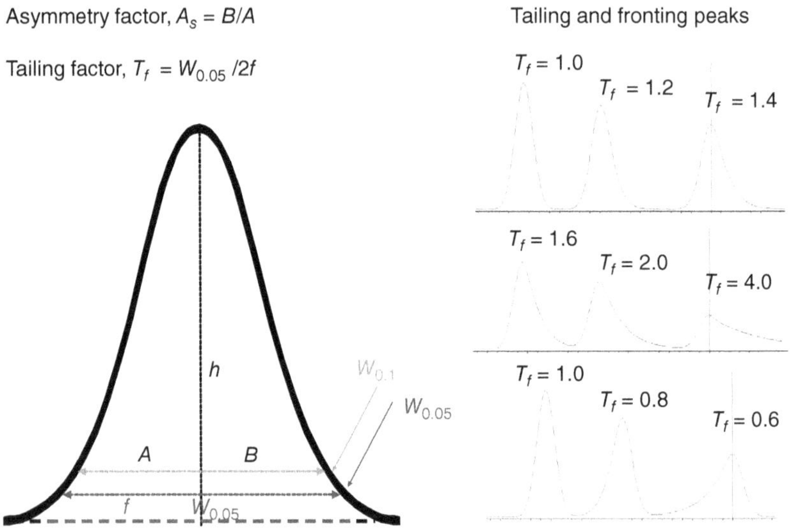

Figure 4.2 A diagram showing the calculation of peak asymmetry (A_s) and tailing factor (T_f) from peak width at 5% height ($W_{0.05}$) according to the USP. Inset diagrams show fronting and tailing peaks.

$$A_s = B / A \tag{4.18}$$

where

B and A = measured at 10% of the peak height

4.3.3 Number of Theoretical Plates (*N*)

Theoretical plate number is a measure of column efficiency and defined by the following equations:

$$N = 16\left(\frac{t_R}{W_b}\right) \tag{4.19}$$

where

t_R = retention time of the analyte peak
W_b = peak width at the baseline

Because of the difficulty in measuring the peak width at the baseline, an alternative equation is used as follows:

$$N = 5.546\left(\frac{t_R}{W_{1/2}}\right)2 \tag{4.20}$$

where

t_R = retention time of the analyte peak
$W_{1/2}$ = peak width at the half height

4.3.4 Capacity Factor (k') or Retention Factor (k)

The capacity factor or retention factor is a measure of the degree of retention of sample components in the column with respect to the void volume, that is, the nonretained peak. It is given as

$$k = \frac{(t_R - t_0)}{(t_0)} \tag{4.21}$$

where

t_R = retention time of the analyte peak
t_0 = retention time of the nonretained peak (It is also called void time [7], hold-up time [7], dead time [10], or breakthrough time [10].)

4.4 Summary

Calculations are integral part of analytical procedures and experimental design to obtain accurate results. Common equations used in routine analytical tests for quantitation of major and/or minor components and impurities in the pharmaceutical products are presented. In addition, common system suitability parameters are discussed. Examples are also included to further demonstrate the applicability of the formulas for both drug substance and drug product. Comprehensive discussion of all pharmaceutical calculations and chromatographic parameters is beyond the scope of this chapter. However, several excellent resources are included in this chapter to provide the readers with more references.

References

1 Davani B, Russo K. A, Wilk A, Bhattacharyya L. Common pharmacopeial calculations in USP monographs. Pharm Forum 31(2) [March–April 2005]; 626–636.
2 Travis P, Heck K, Teitz D, Virgili L, Wiggins M. Compendial Calculations: Improving the calculations within USP–NF. Pharm Forum 31(6) [November–December 2005]; 1749–1755.
3 Kenneth A. Connors, A Textbook of Pharmaceutical Analysis, John Wiley & Sons, Inc., 3rd edition, 2006.

4 James N. Miller and Jane C. Miller, Statistics and Chemometrics for Analytical Chemistry, Pearson Education Limited, 2010.
5 Robert L. Anderson, Practical Statistics for Analytical Chemists, Van Nostrand Reinhold Company Inc., 1987.
6 General Chapter <1010> Analytical Data-Interpretation and Treatment, USP 35 - NF 30, United States Pharmacopeial Convention, Inc., Rockville, MD.
7 Michael W. Dong, Modern HPLC for Practicing Scientists, John Wiley & Sons, Inc., 2006.
8 General Chapter <621> Chromatography, USP 35-NF 30, United States Pharmacopeial Convention, Inc., Rockville, MD.
9 Reviewer Guidance, Validation of Chromatographic Methods, Center for Drug Evaluation and Research (CDER), FDA, 1994.
10 Veronika R. Meyer, Practical High-Performance Liquid Chromatography, John Wiley-VCH Verlag GmbH & Co. KGaA, 2004.

5

Analytical Method Validation, Verification, and Transfer

Behnam Davani

5.1 Introduction

"Analytical method validation" and other similar terms such as "validation of analytical procedure" or "validation of analytical methodology" have been used in the literature, regulatory guidelines, and other resources [1–7]. However, the trend by the pharmaceutical industry has been to adopt the terms and definitions used by the International Conference on Harmonisation of Technical Requirements for Registration of Pharmaceuticals for Human Use (ICH). This is mostly due to their acceptance by the regulatory agencies such as Food and Drug Administration (FDA) in the United States and other regulatory bodies in the European Union, Japan, and other countries.

The validation of analytical procedures (ICH term) is included as part of drug registration applications submitted within the European Union, Japan, and the United States. In addition, such analytical procedure validation is also required for compendial (USP) submissions in support of the development or revision of public standards (monographs) that FDA enforces in the United States.

According to ICH, "the objective of validation of an analytical procedure is to demonstrate that it is suitable for its intended purpose." Other similar definitions are included in the FDA guidance, compendia, and other references. However, the common theme is "suitable for intended purpose or use" or "meet the requirement of the intended analytical application." In other words, the required validation characteristics and limits will depend on the desired specification (tests, procedures, and acceptance criteria) for the intended purpose or application.

5.2 Scope

The validation characteristics used in this chapter are general in nature and apply to most typical and routine methods such as chromatographic

Pharmaceutical Analysis for Small Molecules, First Edition. Edited by Behnam Davani.

(e.g., HPLC and GC) and spectroscopic tests (e.g., IR and UV) used in pharmaceutical laboratories. The validation of specific and more complex tests may need to be handled on a case-by-case basis. In addition, the discussion of the validation parameters and definitions for biological, microbiological, and performance tests (e.g., dissolution) is beyond the scope of this chapter.

5.3 Typical Validation Characteristics

Typical validation characteristics or parameters include the following:

- Accuracy
- Precision
  - Repeatability
  - Intermediate Precision
  - Reproducibility
- Specificity
- Detection Limit (Limit of Detection)
- Quantitation Limit (Limit of Quantitation)
- Linearity
- Range

5.4 Definition and Determination of Analytical Characteristics

5.4.1 Accuracy

The accuracy of an analytical procedure is the closeness of the results found to an acceptable true value or a reference value.

For the assay of a drug substance, this can be performed by application of the procedure to a reference standard. If the reference standard is not available, comparison of the results from the procedure to those obtained from another validated and well-characterized analytical procedure can be made.

For the assay of a drug product, accuracy can be determined by application of the procedure to synthetic mixtures of the drug product components spiked with known quantity of the drug substance. Alternatively, the procedure can be applied to the drug product spiked with known quantities of the analyte. Another option is to compare the procedure results to those obtained from a well-characterized or recognized (official) procedure.

For quantitative analysis of impurities, accuracy can be assessed by spiking the drug substance or product with known quantities of the impurities and evaluating the analyte recoveries. Alternatively, the accuracy can be determined by comparison of the results using the analytical procedure to those from another well-characterized or official procedure.

Table 5.1 Accuracy data for assay using spike recovery method.

Percent of labeled claim	Percent recovery (sample preparation 1)	Percent recovery (sample preparation 2)
80	98.1	98.2
90	101.0	98.1
100	100.6	100.3
110	100.4	101.1
120	100.7	100.8
Average ($n = 5$)	100.2	99.7
% RSD	1.2	1.4
Range ($n = 5$)	98.1–101.0%	98.1–101.1%
Acceptance criteria:		
Individual recovery = 97.5–102.5%		
Average recovery = 98.0–102.0%		
% RSD ≤ 2		

An example of typical accuracy data and acceptance criteria is given in Table 5.1. Note that this experimental design can also be used to measure the variation (precision) of the procedure.

5.4.2 Precision

The precision of an analytical procedure is the degree of agreement obtained between individual results from multiple sampling of the same homogeneous sample.

5.4.2.1 Repeatability

The precision obtained by one analyst under the same operating conditions in a short period of time.

For the assay or quantitative determination of impurities, precision is determined by analysis of multiple sample preparations. This precision is typically expressed as standard deviation or relative standard deviation (coefficient of variation). According to ICH recommendation, this should be done upon a minimum of nine determinations within the specified range or upon a minimum of six analyses at 100% of the sample concentration.

5.4.2.2 Intermediate Precision (Ruggedness)

The precision obtained by different analysts, different equipment, on different days, and so on, within the same laboratory.

Table 5.2 Precision data for assay (repeatability and intermediate).

Sample preparation number	% Result (analyst 1 day 1 instrument 1)	% Result (analyst 2 day 2 instrument 2)	Difference
1	99.8	98.5	0.3
2	100.4	98.0	2.4
3	99.7	98.5	1.2
4	100.0	101.5	1.5
5	100.5	98.3	1.2
6	99.8	98.1	1.7
Average ($n = 6$)	100.0	98.8	1.4
%RSD ($n = 6$)	0.3	1.3	
Acceptance criteria: %RSD ≤ 2 and average of differences ≤ 2			

The effects of different and normal variations on the analytical results within the same laboratory need to be evaluated. Meaningful acceptance criteria based on the analytical procedure and its intended use should be established. Statistical approaches such as *F*-test and other more complex statistical techniques can also be used to assess the significance of differences of the results obtained under these conditions [8].

An example of typical precision data and acceptance criteria is given in Table 5.2. In this experimental design, both repeatability and intermediate (ruggedness) precision can be evaluated using percent relative standard deviation and percent difference of averages, respectively.

5.4.2.3 Reproducibility

The precision obtained between different laboratories at different sites (e.g., collaborative studies).

This precision is used for standardization of the procedure and involves collaborative studies among multiple laboratories at different sites. The extent of study and the number of laboratories selected depend on many factors including the complexity of the procedure and the intended use of the procedure. The precision from reproducibility study is usually lower (larger relative standard deviation) than those obtained from the ruggedness studies. This is expected because there is a greater variability of the experimental conditions and staff training among multiple laboratories.

5.4.3 Specificity

The specificity is defined by ICH guideline as "the ability to assess unequivocally the analyte in the presence of components expected to be present." These components include impurities, degradation products, matrix, and others. Specificity has also been defined as the ability of the procedure to discriminate between the analyte(s) of interest and other components such as expected or potential impurities in the sample.

This determination can be performed by analysis of samples containing the analyte and to confirm the positive result using a reference standard. In addition, the negative result can be confirmed by applying the procedure to the samples known not to contain the analyte.

For the assay, the specificity can be demonstrated by separation or discrimination of the analyte of interest from other expected components and/or impurities. This can be achieved by spiking the sample with known amount of such impurities and demonstrating that the assay result is not affected.

For the impurities, this can be accomplished by spiking the drug substance or drug product with known quantities of impurities and demonstrating that these are separated (resolved) from the major compound (drug substance). If impurities and reference standards are not available, the specificity can be shown by comparing the results to those obtained from another well-characterized or official procedure.

It is ideal to use a single test for identification of the analyte. However, it may not be practical to have a single test for positive identification. In this case, orthogonal (complementary) tests such as IR (spectral match) and HPLC (retention time match) are commonly used to identify the analyte. On the other hand, an HPLC with UV detection and a UV test are not orthogonal tests and thus are not considered sufficiently specific for positive identification.

5.4.4 Detection Limit (DL)

The lowest amount or concentration of the analyte in the sample, which can be detected but not necessarily quantitated.

For noninstrumental procedure, it is estimated from the lowest concentration of the analyte that can be detected but not necessarily quantitated. For instrumental procedure, it is estimated from the concentration of analyte with a signal-to-noise ratio of typically 3:1.

Another approach for calculation of the detection limit is using the following equation [1, 9]:

$$\mathrm{DL} = 3.3\ s/b$$

where

s = standard deviation of the background response for the blanks or for analyte response at low concentration levels

b = slope of the calibration curve for the analyte

5.4.5 Quantitation Limit (QL)

The lowest amount or concentration of the analyte in the sample, which can be quantitated with suitable level of precision and accuracy.

For noninstrumental procedure, it is the lowest concentration of the analyte that can be quantified under experimental condition with acceptable precision and accuracy. For instrumental procedure, it is estimated from the concentration of analyte with a signal-to-noise ratio of 10:1.

Another approach for calculation of the quantitation limit is using the following equation [1, 9]:

$$\text{QL} = 10\ s / b$$

where

s = standard deviation of the background response for the blanks or analyte response at low concentration levels

b = slope of the calibration curve for the analyte

It is important to note that the detection and quantitation limits using the aforementioned formulas are only estimated values. Therefore, the actual values need to be experimentally verified. Furthermore, QL is estimated to be about three times larger than DL based on the aforementioned formulas.

An example of typical detection and quantitation limits is given in Table 5.3. The selected data points should encompass the lower concentration range established during the linearity study [9].

Table 5.3 Detection and quantitation limits for impurities.

Replicates	Response (1 μg/mL)	Response (2 μ/mL)	Response (3 μg/mL)
1	3020	6011	9011
2	2980	5989	8798
3	2999	6112	8908
4	3045	5977	9002
5	2989	6056	8945
6	3011	6009	8969
Average ($n = 6$)	3007	6025	8939
Standard deviation (SD)	23	50	79
Average standard deviation	51		
Slope of calibration curve	2979		
Detection limit	$3.3 \times 51/2979 = 0.06\ \mu\text{g/mL}$		
Quantitation limit	$10 \times 51/2979 = 0.2\ \mu\text{g/mL}$		

Detection and quantitation limits are more commonly expressed as percent of the active ingredient rather than concentration unit (e.g., μg/mL). However, this unit can be converted to percent by dividing this value by the concentration of the test solution containing the active ingredient shown as follows:

Assume: 200 μg/mL of test solution

$$\mathrm{DL} = 0.06\ \mu\mathrm{g/mL} / 200\ \ \mu\mathrm{g/mL} \times 100 = 0.03\%$$
$$\mathrm{QL} = 0.2\ \mu\mathrm{g/mL} / 200\ \ \mu\mathrm{g/mL} \times 100 = 0.1\%$$

5.4.6 Linearity

The ability of an analytical procedure to produce a response that is directly or through mathematical transformation (e.g., log, square root) proportional to the concentration (amount) of the analyte in the sample within a given range.

The linear relationship between the analyte response and the corresponding concentration is evaluated by statistical or mathematical approach. One common procedure is the generation of regression plot using the least squares method and calculation of the correlation coefficient (r). Correlation coefficient is one measure of degree of linear relationship between two variables [10]. The correlation coefficient values range from −1 to +1. The −1 value indicates perfect negative correlation, and +1 indicates perfect positive correlation. When $r = 0$, there is no linear correlation or association. For the establishment of linearity, minimum of five concentration levels are recommended by ICH. The correlation coefficient of 0.99+ is typically desired. Lower values can be used and justified for certain analyses or specific applications. The intercept is a measure of the bias of the procedure, and thus, it should be close to zero (statistically insignificant).

An example of linearity data for impurities is highlighted in Table 5.4. The 100% target refers to the concentration corresponding to acceptable impurity limit usually expressed in percentage of the active ingredient. In this example, assuming an acceptable impurity limit of 0.1%, the range from 50% to 150% of the target concentration corresponds to impurity levels at 0.05%, 0.075%, 0.1%, 0.125%, and 0.15% as shown in column 3 of Table 5.4.

5.4.7 Range

The interval between the upper and lower concentration levels (including these levels) of the analyte in the sample demonstrated to show suitable level of precision, accuracy, and linearity.

For the assay of the drug substance or drug product, test concentrations of 80–120% are recommended. For the determination of impurities, a range of 50–120% of the acceptable impurity limit is generally used. For other types of analyses, different ranges may be considered and justified.

Table 5.4 Linearity data for impurities.

Target concentration	Analyte concentration (μg/mL)	Analyte concentration (% of target)	Response
QL	0.1	0.025	9,890
50%	0.2	0.05	20,759
75%	0.3	0.075	30,301
100%	0.4 (target)	0.1 (target)	40,399
125%	0.5	0.125	46,800
150%	0.6	0.15	60,764
Slope	97,792		
Intercept	326		
Correlation coefficient	0.99700		

5.5 Types of Analytical Procedures

- Identification test
- Quantitative test (assay) for major component(s) in the drug substance or drug product
- Quantitative test for impurities
- Limit test for impurities

5.6 Typical Validation Requirement

The list of validation characteristics considered most important for the validation of different types of analytical procedures is included in Table 5.5.

The robustness has not been included in this list as part of the analytical procedure validation. However, this characteristic should be evaluated as part of the development of the procedure to examine the effect of small but deliberate variations of critical parameters on the results. In addition, a ruggedness study (beyond the intermediate precision) is recommended to assess the effect of operational and environmental variables on the test results. In this case, a second analyst repeats the entire validation parameters using different instruments on different days to further verify the reliability of the procedure in meeting the required acceptance criteria established before the start of study.

It is also important to note that it may not be possible or practical in some cases to use one analytical procedure to meet all the validation characteristics.

Table 5.5 Validation characteristics for different types of analytical procedures.

Analytical procedure characteristics	Identification	Quantitative test (assay)	Quantitative test for impurities	Limit test for impurities
Accuracy	No	Yes	Yes	No
Precision	No	Yes	Yes	No
Specificity	Yes	Yes	Yes	Yes
Detection limit	No	No	No[a]	Yes
Quantitation limit	No	No	Yes	No
Linearity	No	Yes	Yes	No
Range	No	Yes	Yes	No

Source: Table adapted from ICH Harmonised Tripartite Guideline Q2 (R1).
a) May be needed in some cases.

In such cases, more than one procedure can be used to meet all the required characteristics and acceptance criteria.

5.7 Revalidation

It may be necessary to revalidate the procedure depending on several factors including the following [1, 2]:

- Different routes of synthesis resulting in different impurity profiles or other characteristics (e.g., different solubility, water of hydration) for the drug substance
- Difference in the composition of the drug product due to different formulation and matrix
- Major changes in the analytical procedure such as substitution of titration or TLC with HPLC procedure.

The degree of revalidation depends on the extent of the change in the analytical procedures. Simple and single change to noncritical parameters may not require revalidation. The robustness study performed in the development of the analytical procedure provides valuable information to assess the significance of the change and the need for revalidation.

5.8 System Suitability

System suitability test (SST) and requirements are considered integral part of the analytical procedures for pharmaceutical analysis [11, 12]. The SST applies to both chromatographic and nonchromatographic (e.g., spectroscopic)

Table 5.6 System suitability parameters and typical criteria.

Parameter	Typical acceptance criteria
Percent relative standard deviation (% RSD)	≤2 for major component 5%–15% for trace components
Resolution	>2.0 in General
Tailing factor	≤2
Number of theoretical plates	>2000
Limit of detection	0.03%
Limit of quantitation	0.05%
Capacity factor	>2

methods. The SST is to ensure that the system (equipment, electronics, operator, samples, and other environmental factors) functions properly at the time of analysis. However, SST should not be considered as a substitute for quality systems such as calibration and equipment qualifications.

The SST and requirements depend on the type of sample (drug substance or dosage form), analytical procedure, and required specifications for the intended use. Therefore, the experimental design and specifications need to be carefully established before the development and validation of procedures. Some typical parameters and desired acceptance criteria for chromatographic procedures are included in Table 5.6 [4].

Not all the system suitability parameters may be needed or critical for specific procedure. For instance, there is no need to determine the limit of quantitation for a major component in the drug substance (See Table 5.5). In addition, resolution is generally considered a better measure of the degree of separation of components than the number of theoretical plates. Therefore, these parameters and desired criteria must be carefully evaluated for the intended purpose of the analysis and/or regulatory requirements.

5.9 Forced Degradation (Stressed) Studies

There are several objectives for performing forced degradation studies [13–18]. These include obtaining relevant information about potential degradation products, degradation pathway, and product stability. In addition, such information helps the analyst to evaluate the specificity of an analytical procedure. Such procedure is also called stability-indicating assay because it is validated to accurately determine the active ingredient in the presence of potential degradation products and impurities.

Table 5.7 Stress parameters and typical conditions for drug substance.

Parameter	Conditions (as solid)	Conditions (as solution or suspension)
Hydrolysis (acid, base, and thermal)	–	At different pH ranges using 0.1–1 N HCl or 0.1–1 N NaOH
Oxidation	–	H_2O_2 (protected from light)
Photodegradation	Fluorescent and UV light	Fluorescent and UV light
Thermal	50–70 °C	–
Humidity	75%	–
Thermal/humidity	50–70 °C/75%	–

Table 5.8 Stress parameters and typical conditions for drug product.

Parameter	Conditions (as solid)	Conditions (as liquid)
Hydrolysis (acid, base, and thermal)	–	At different pH ranges using 0.1–1 N HCl or 0.1–1 N NaOH
Oxidation	–	H_2O_2 (protected from light)
Photodegradation	Fluorescent and UV light	Fluorescent and UV light
Thermal	50–70 °C	–
Humidity	75%	–
Thermal/humidity	50–70 °C/75%	–

The stress studies are usually performed on one lot (batch) of the drug substance or product. The stress conditions are designed to provide partial (about 5–20%) degradation. The intent is to evaluate the impact of reasonable worst-case environmental conditions (e.g., high temperature, humidity, light exposure) rather than complete decomposition of the samples.

The parameters and typical stress conditions for drug substances and drug products are highlighted in Tables 5.7 and 5.8, respectively [14, 19].

5.10 Analytical Method Verification

Compendial tests are considered validated, and therefore, there is no requirement to revalidate these procedures. However, according to the current good manufacturing practices (GMPs), FDA requires that suitability of all testing methods to be verified under actual condition of use. Simple compendial wet chemistry tests such as pH and loss on drying may not require verification. However, more complex procedures such as chromatography, spectroscopy, titration require method verification [20].

Validation challenges the analytical method using a well-defined sample. On the other hand, verification method challenges the analytical environment using a well-defined method (compendial). The extent of assessment for verification depends on multiple factors including the following:

1) Analyst (education, training, experience)
2) Instrument
3) Reagents
4) Matrix

The first three aforementioned factors are important and part of GMP requirement. In other words, laboratories are expected to have trained analysts using qualified instruments with appropriate purity of reagents. From this point of view, the most critical (unknown) is the sample matrix because different formulations containing different excipients may provide analytical challenges compared to compendial procedure. More importantly, the drug substance or product may have completely different impurity profiles due to different synthetic route and manufacturing process. Thus, the compendial procedure may not work for such products. This is indeed the most important reason for verification of the compendial procedures to determine the suitability of the procedures to drugs manufactured with different formulations and processes. One needs to keep in mind that the compendial procedures are only suitable for their intended use (generally approved specification by FDA). In addition to potential impurity profile, other attributes such as water of hydration, solvent, polymorphism may be different from those in compendial procedures. When applicable, *USP* monographs provide multiple procedures (e.g., different procedures for impurities, dissolution) through flexible monograph approach for articles (drug substances and products) with different impurity profiles, dissolution, or other tests as appropriate.

In general, selective validation parameters are evaluated to verify the compendial procedures. However, there is no specific guideline about what parameters to be performed. Meeting the system suitability criteria is the first step. Then, specificity, accuracy, and precision are the most important criteria. The linearity also needs to be assessed in cases where different instruments have different linearity range. One has to keep in mind that where verification criteria are specified in individual USP monographs or in mandatory general chapters (numbered below 1000), these criteria will take precedence over the guidelines in the information general chapter <1226> verification.

5.11 Analytical Method Transfer

The objective of this study is to ensure that the receiving lab obtains the same (equivalent) results as those obtained by the initiating (transferring) lab [21]. This can be accomplished by several approaches outlined as follows.

5.11.1 Comparative Testing

This is the most common approach and involves testing of multiple lots (typically three if available for quantitative analysis) and comparing the test results. The acceptance criteria for result differences are based on the type of test (assay, impurity, dissolution, etc.), whether the test is quantitative or qualitative, sample/matrix, and most importantly, the intended use and requirements. In general, a lower absolute difference (1–2%) is expected for the assay of drug substance/product than for the impurities (about 5–20% based on the level of impurities). Similar relative standard deviations are desired for assay and impurities. However, further discussion of result equivalency requires more comprehensive statistical analysis and is beyond the scope of this chapter. The readers are advised to consult with the company's statistician and other references [22–24] for experimental design and the desired level of confidence in establishing the acceptance criteria.

5.11.2 Co-Validation between Labs

The validation is completed typically by the originating lab before the method transfer. However, in this approach, the receiving lab can become qualified by concurrently performing some aspects of the validation characteristics, especially intermediate precision.

5.11.3 Revalidation

Revalidation or partial validation by the receiving lab is another approach for qualification of the lab. One example is when the initial validation is outdated and better instrumentation/requirements are needed.

5.11.4 Transfer Waiver

It is possible to waive the transfer for the receiving lab under special circumstances. Some examples of such circumstances include the following:

- The personnel performing the test in the receiving lab are the same as those who actually performed the initial validation in the transferring lab. This can happen with merger and/or acquisition of the lab by other labs/companies. In this case, the transferring lab is basically changing the name, but the staff and infrastructure/equipment stay the same.

- The staff performing the test was transferred to the receiving lab.
- A similar procedure including instrumentation has been routinely used in the receiving lab for the same/similar product.

The elements recommended for method transfer include the following:

- Preapproved test protocol including transfer acceptable criteria
- Description of analytical procedure
- Transfer report

A detailed protocol, procedure, and criteria are critical to minimize and handle out-of-specification results. The protocol and procedure should also include the required purity of reference standard(s) or reagents as well as specific instrumentation needed to successfully perform the method transfer.

5.12 Summary and Conclusion

Reliable analytical results are necessary to make informed decision about the quality and safety of the products in the pharmaceutical industry. In addition, such analytical data are required for regulatory submissions in support of the drug product registrations. Therefore, meaningful experimental designs including system suitability parameters must be planned for the intended use of the procedure. In this chapter, general guideline for the determination of the analytical characteristics for different types of validation procedures was highlighted for the analysis of both the drug substance and drug product. The factors to consider for verification of the compendial procedures were discussed. In addition, different approaches for the transfer of analytical procedure from one lab (transferring) to other lab(s) (receiving) under different circumstances were described.

References

1 ICH Harmonised Tripartite Guideline, Q2 (R1), Valuation of Analytical Procedures: Text and Methodology, November 1996.
2 Guidance for Industry: Analytical Procedures and Methods Validation; Chemistry, Manufacturing, and Controls Documentation, Draft August 2000; U.S. Department of Health and Human Services, Food and Drug Administration, Center for Drug Evaluation and Research (CDER), Center for Biologics Evaluation and Research (CBER).
3 General Chapter <1225> Validation of Compendial Procedures, USP 38-NF 33, United States Pharmacopeial Convention, Inc., Rockville, MD.
4 Reviewer Guidance, Validation of Chromatographic Methods, Center for Drug Evaluation and Research (CDER), FDA, November 1994

5 Analytical Chemistry in a GMP Environment, A Practical Guide, edited by James M. Miller and Jonathan B. Crowther, John Wiley & Sons, Inc., 2000.
6 Analytical Method Validation: Back to Basics, Part I, by Michael Swatz and Ira Krull, LCGC North America, Nov. 1, 2009.
7 Analytical Method Validation: Back to Basics, Part II, by Michael Swatz and Ira Krull, LCGC North America, Jan. 1, 2011.
8 James N. Miller and Jane C. Miller, Statistics and Chemometrics for Analytical Chemistry, Pearson Education Limited, (2010).
9 Analytical Method Validation, Journal of Validation Technology, Special Edition, Volume 1, 2006, Advanstar Communications Inc.
10 Practical Statistics for Analytical Chemists, Robert L. Anderson, Van Nostrand Reinhold Company Inc., 1987.
11 General Chapter <621> Chromatography, USP 38-NF 33, United States Pharmacopeial Convention, Inc., Rockville, MD.
12 Method Validation in Pharmaceutical Analysis, A guide to Best Practice, Edited by Joachim Ermer, John H. McB. Miller, Wiley-VCH Verlag GmbH@ Co. KGaA, 2005.
13 ICH Q1 A, Stability testing of new drug substances and products, 2003.
14 ICH Q1B Guideline, Photostability Testing of New Drug Substance and Products, 1996.
15 Thatcher, S. R., Mansfield, R. K., Miller, R. B., Davis, C. W., and Baertschi, S. W., Pharmaceutical Photostability: A Technical Guide and practical Interpretation of the ICH Guideline and Its Application to Pharmaceutical Stability–Part 1, Pharmaceutical Technology, (2001) pp. 98–110.
16 Thatcher, S. R., Mansfield, R. K., Miller, R. B., Davis, C. W., and Baertschi, S. W., Pharmaceutical Photostability: A Technical Guide and Practical Interpretation of the ICH Guideline and Its Application to Pharmaceutical Stability–Part II, Pharmaceutical Technology, (2001), pp. 50–62.
17 Reynolds D, Forced degradation of pharmaceuticals. Am Pharm Rev 7 [2004]; 56–61.
18 Dolan J. W., Stability-indicating assay. LCGC N Am 20 [2002]; 346–349
19 ICH Q1 A (R2), Stability testing of new drug substances and products, 2003.
20 General Chapter <1226> Verification of Compendial Procedures, USP 38-NF 33, United States Pharmacopeial Convention, Inc., Rockville, MD.
21 General Chapter <1224> Transfer of Analytical Procedures, USP 38-NF 33, United States Pharmacopeial Convention, Inc., Rockville, MD.
22 Statistics and Chemometrics for Analytical Chemistry, James N. Miller and Jane C. Miller, sixth edition, Pearson Education Limited, 2010.
23 ISPE, Good Practice Guide: Technology Transfer, Analytical Methods, ISPE, (2003), pp. 23–24.
24 Swartz M., Krull I., Analytical method transfer. LCGC N Am 24(11) [Nov. 2006]; 1204–1214.

6

Specifications

Ernest Parente

6.1 Scope

This chapter is primarily focused on establishing specifications for small-molecule drug substances and drug products. The chapter has been developed using the principles and recommendations in the ICH[1] guideline Q6A, "Specifications: Test Procedures and Acceptance Criteria for New Drug Substances and New Drug Products: Chemical Substances" [1], for the selection of tests and procedures. This guideline is an internationally recognized standard for developing specifications for drug substances and drug products. In addition to the recommendations for tests and procedures, the chapter presents several approaches to establishing acceptance criteria for tests commonly included in drug substance and drug product specifications. Although the scope of the ICH guideline is focused on requirements for new product applications, the principles in the guideline can be applied to marketed products. In addition, while the concepts described in the ICH guideline address the specifications for drug substances and drug products, they can easily be extended to include inactive formulation ingredients. For biologic therapeutics, the reader is referred to ICH guideline Q6B [2], which uses a similar approach but focuses on special considerations for specifications for biotechnological/biological products.

6.2 Introduction

According to ICH guideline Q6A, "A specification is defined as a list of tests, references to analytical procedures, and appropriate acceptance criteria which are numerical limits, ranges, or other criteria for the tests described. It establishes the set of criteria to which a new drug substance or new drug product should conform to be considered acceptable for its intended use" and should

1 International Conference on Harmonization (ICH).

Pharmaceutical Analysis for Small Molecules, First Edition. Edited by Behnam Davani.

ensure that a substance or drug product meets the standards of identity, strength,[2] quality, and purity throughout its shelf life.

The acceptance criteria can be expressed as a quantitative limit, a range of values, or a qualitative requirement. The guideline further recommends "Universal Tests" that must be included in all specifications and "Specific Tests" that are related to quality attributes or performance characteristic of a given drug substance or drug dosage form. The rationale for the selection of tests, procedures for testing, and acceptance criteria should be included as part of the justification for specifications. In some cases, flow diagrams called "decision trees" are provided in the ICH Q6A guideline to aid in the selection of appropriate tests. However, it is ultimately the responsibility of the product manufacturer to provide a complete justification of specifications.

An example of a specification for "Universal" tests for a hypothetical drug substance is given in Table 6.1.

Table 6.1 Example of specifications for universal tests for a drug substance.

Test	Acceptance criteria	Procedure reference
Description	White to off-white powder	M1234
Identification		
A. Retention time	Exhibits the same retention time as reference standard	M2345 (HPLC)
B. Infrared spectroscopy	Exhibits the same maxima and minima as a reference standard similarly prepared	USP <197K>[a]
Assay	98.0–102.0%	M0123 (HPLC)
Impurities		
Organic impurities		M0123 (HPLC)
Impurity A	NMT 0.20%	
Impurity B	NMT 0.30%	
Unspecified impurities	NMT 0.10%	
Total impurities	NMT 0.50%	
Inorganic impurities	NMT 0.10%	USP <281> (residue on ignition)
Residual solvents		USP <467> (residual solvents)
Acetonitrile	NMT 410 ppm	

HPLC, high-performance liquid chromatography; NMT, not more than.

a) Refer to USP General Chapter – Spectrophotometric Identification Tests

2 The terminology "strength" is preferred for chemical medicines, while "potency" is usually used for biological materials.

The table shows the format for preparing a specification using the ICH recommendation including the test, acceptance criteria, and procedure reference. The table gives examples of tests with qualitative and/or semiquantitative acceptance criteria (e.g., retention time and IR spectroscopy), acceptance criteria specifying a quantitative range (e.g., assay), and acceptance criteria limits for quantitative tests (e.g., organic impurities). In addition, note that the specification reference methods developed as "in-house" methods, designated as "M" method tests, and references to procedures are provided in General Chapters in the United States Pharmacopeia (USP) compendium [3], which is an excellent source of information concerning tests and procedures. A complete set of specifications would also include relevant specific tests and performance-related tests. For examples of typical specifications for a drug substance, an oral tablet, and an oral solution, the reader is referred to the chapter by Riley and Little [4].

6.3 Types of Tests

Following is a list of typical types of tests selected for developing a specification. Selection of the test format depends on the nature of the attribute being tested and the method of testing being employed.

1) Limit Tests are qualitative or semiquantitative tests that are usually used to control low-level impurities. These tests have "Pass/Fail" acceptance criteria. For example, the limit for chloride could be controlled by comparing the opalescence of a test solution to that of a standard of known concentration prepared at the specified limit.
2) Qualitative Tests are nonnumerical or categorical tests that are conducted to ascertain a qualitative aspect of a given attribute. Many qualitative tests are conducted by visual examination (e.g., appearance tests or visual comparisons, infrared spectroscopic identity). There is an increasing trend of replacing tests that have qualitative end points with quantitative tests that are less subjective.
3) Quantitative Tests are tests with numerical end points that are compared to limits or ranges specified by numerical acceptance criteria. Drug product assays and impurity tests conducted by high-performance liquid chromatography (HPLC) are examples of quantitative test.
4) In-Process Tests are conducted during a manufacturing process to indicate that the process has achieved a given acceptance criteria at a given manufacturing stage. If the material passes the in-process test, it is released for further processing. Otherwise, it may need to be reprocessed, reworked, or rejected. Test results and acceptance criteria for in-process tests are usually included in a manufacturing batch record. However, they can also be

controlled by a separate in-process specification. In-process tests can be conducted on-line or off-line and can involve real-time feedback or feedforward processing. An off-line test generally involves sending a sample to a remote location (i.e., testing laboratory) for testing. An example of an on-line, in-process test is the continuous and automatic adjustment of the compression force of a tablet press to product tablets of a specified hardness value. The advent of Process Analytical Technology (PAT), the use of Quality by Design (QbD) concepts, and establishing acceptable "Design Spaces" for acceptance or rejection of quality attributes are becoming more common in pharmaceutical manufacturing processes [5].

5) Parametric Tests measure parameters that are indicative of compliance in lieu of directly measuring a quality attribute. An example of this is the measurement of the sterilization time and temperature in a sterilization process. While the actual measurement of sterility by biological testing is the key attribute, achieving validated specification for sterilization time and temperature parameters is indicative of achieving the specification. "Parametric release" is a concept recognized in ICH guideline Q6A.
6) Periodic Quality Indicator Tests (PQIT), also called "skip tests," are tests included in a specification that are not tested on every batch but are tested periodically according to a given schedule (e.g., tested every 10th batch) to monitor compliance. This type of testing is often used for periodically monitoring a given quality attribute for raw materials that are tested under a reduced testing protocol. A typical example would be the periodic testing of microbial attributes for solid oral dosage forms shown to be at low risk for microbial bioburden. PQIT should only be used for quality attributes with a very low safety risk and a low risk of failure since a failed result will call into question the quality of all lots released since the last successful test.

6.4 Types of Specifications

Following is a brief discussion of general types of specifications commonly used for drug substances, drug products, and excipients of pharmaceutical interest. Additional information for related concepts and terminology is also included.

Release Specifications. Release specifications are a list of tests, procedures, and acceptance criteria that need to be met for the release of a product or material for distribution. The tests need to adequately characterize the critical quality attributes of the material or drug product, and the acceptance criteria need to be established so that the test article will remain within specifications over its shelf life. Release specifications are directly related to the shelf-life specifications and the proposed expiration date. The release specifications are generally derived from process capability considerations, and the process needs to be designed to produce products of adequate quality. In the European Union (EU), release specifications are a regulatory requirement for a marketing authorization. In the United States (US) and Japan, release specifications are generally not specifically required for product registration, but are usually controlled as "in-house" release criteria to ensure that quality standards are met at the time of manufacture. However, in some cases where there is a significant change in a critical quality attribute with time, a release specification may be useful or required to justify a shelf-life proposal.

Shelf-Life Specification. Shelf-life specifications are a list of tests, procedures, and acceptance criteria that need to be met. If met, these criteria usually provide at least 95% confidence that the product will remain within registration specifications throughout the shelf life. It needs to reflect the process capability, stability profile, measurement precision and justify the proposed product expiry period. The tests selected need to address the critical quality attributes (CQA) for a given material or drug product dosage form. Shelf-life specifications are legal requirements, and the failure of a product in distribution to meet a shelf-life specification results in the withdrawal of the product from the market.

Interim Specification. An interim specification is a provisional, but legally binding, regulatory specification that is used to control a given quality attribute during a period in which the proposed specification is evaluated. Interim specifications for a given attribute can be negotiated with a competent regulatory authority (e.g., Food and Drug Administration, FDA) when insufficient data are available at the time of filing with a postapproval agreement to reevaluate the specification once additional process capability data or stability data become available.

"Sunset" Specifications. These are provisional specifications that are in force for a specified period of time that will eventually be deleted from the final specifications if certain requirements are met. For example, it could be possible to "sunset" a test after satisfactory completion of a stability program or the manufacture of a sufficient number of full-scale batches to assess the capability of a process. The use of a sunset strategy typically requires regulatory preapproval.

Following are the definitions for other terms related to material or product specifications:

Expiration Date. This is the date after which a material can no longer be used or distributed (also called the discard date). The expiration date should be justified by the data demonstrating that the material or product is fit for its intended use. Discard dates for some materials, reagents, solutions, and so on can be set if sufficient information is available regarding their stability. In cases where the expiration date is expressed as a month and a year (e.g., June 2020), standard operating practices usually allow the material to be used until the last day of the month. One should be aware that there are different conventions for expressing dates in different jurisdictions. For example, 6/12/2020 refers to June 12, 2020, by US convention, but would be interpreted as December 6, 2020, by European convention.

Recontrol Date or Retest Date. This is the date after which a material needs to be retested to reconfirm its fitness for use. Typically, raw materials and reference standards are controlled using recontrol testing in which all or parts of the quality specifications are reconfirmed to show that the material is suitable for its intended use. For example, after a specified time, quality attributes of a drug substance that may be expected to change with time (e.g., impurities, water content) may be reassessed as part of a recontrol procedure, and if suitable, its use can be extended. The recontrol period needs to be justified by data and the number of times the shelf life of a material can be extended by a recontrol procedure should be specified to determine the expiration date. Expiration and recontrol dates do not necessarily mean that the item has degraded, but that the length of time an item is suitable for its intended purpose is determined by the extent and duration of the stability study, and that generally the industry does not elect to study stability greater than 5 years and often no more than 2 years. For inexpensive excipient ingredients or laboratory chemicals, it may be more cost-effective to discard and replace the material than to conduct retesting.

Hold Time. This is the length of time that a material can be held in a temporary storage container after which it would need to be recontrolled to demonstrate fitness for use. Usually, in-process materials or finished drug products waiting to be packaged are assigned permitted hold times. Similarly to expiration dates and recontrol dates, hold times need to be supported by data, particularly, for extremely long (usually >1 month) hold times that could occur for drug products being transported to a secondary packaging site. For materials that exceed the hold time, they are usually recontrolled to ensure that they meet the material release specification with the assumption that after packaging, the product will still meet the established product expiration date.

Beyond Use Date. This is an "expiration date" assigned to compounded preparation after which it should not be used, dispensed, or stored. The beyond use date is usually assigned to a compounded drug product preparation based on the date or time the preparation was compounded.

6.5 Selection of Tests and Procedures

6.5.1 Universal Tests

The four universal tests required in the specifications for all new drug substances and drug products include the following: description, identification, assay, and impurities. Following are the definitions for each of the universal tests as provided in the ICH Q6A guideline [1].

6.5.1.1 Drug Substances

The following tests and acceptance criteria are considered generally applicable to all new drug substances.

a) Description: "A qualitative statement about the state (e.g. solid, liquid) and color of the new drug substance. If any of these characteristics change during storage, this change should be investigated and appropriate action taken." With regard to color, if changes are observed over time, an attempt should be made to use a quantitative measure to describe the color change. Several instruments[3] are available that quantify color using a color space model such as the CIE[4] L*a*b* (CIELAB) tristimulus model [6]. While the instrumental technique removes the subjectivity of the observer and simplifies the communication of the color change, a qualitative judgment of what is or is not an acceptable color difference is still needed.
b) Identification: "Identification testing should optimally be able to discriminate between compounds of closely related structure which are likely to be present. Identification tests should be specific for the new drug substance, e.g., infrared spectroscopy. Identification solely by a single chromatographic retention time, for example, is not regarded as being specific. However, the use of two chromatographic procedures, where the separation is based on different principles or a combination of tests into a single procedure, such as HPLC/UV diode array, HPLC/MS, or GC/MS is generally acceptable. If the new drug substance is a salt, identification testing should be specific for the counterion. An identification test that is specific for the salt itself should suffice. Drug Substances which are optically active may also need specific identification testing, such as optical rotation or performance of a chiral assay."

   To establish identity, spectroscopic, chromatographic, and chemical tests are typically employed. A list of common spectroscopic and chromatographic tests used to establish identity is provided in Table 6.2.

   Usually, two identification tests that employ different principles for discrimination (so-called orthogonal test) are used. Because of its specificity and ability to discriminate even closely related compounds, infrared spectroscopy is considered the method of choice. The test article is generally compared to a reference standard of known identity, and the identification is considered positive if the sample exhibits absorption maxima at the same wavelengths as the standard. While the presence of additional maxima

3 Suitable tristimulus colorimeters are available from Hunter Associate Laboratories, Inc., Reston, VA or BYK-Gardner USA, Silver Spring, MD.
4 Commission Internationale de L'Eclairage (CIE).

Table 6.2 Common spectroscopic and chromatographic tests used for identification testing.

Spectroscopic tests	Chromatographic tests
Infrared spectroscopy	High-performance liquid chromatography (HPLC)
Ultraviolet and visible spectroscopy	Gas chromatography
Near-infrared spectroscopy	Thin-layer chromatography
Raman spectroscopy	Chiral chromatography
Nuclear magnetic resonance spectroscopy	
Mass spectrometry	
X-ray diffraction	
Optical rotation	

wavelengths can sometimes be attributed to the presence of impurities, the absence of an absorption band is evidence of a negative identification. While not considered as specific as infrared spectroscopy, comparison of the chromatographic retention of a test article to the retention of a reference standard of known identity is commonly used as a second method of identification. In addition to being an "orthogonal" method, it has the added advantage that it does not require additional work to conduct the testing since HPLC is often used for assay or impurity testing.

An identification test for the counter ions, if present, should also be included (i.e., chloride test for a drug supplied as a hydrochloride salt). With regard to counter-ion identification, a specific, qualitative test, usually performed by a wet-chemical procedure, is typically used. However, quantitative tests may be useful in establishing identity and purity. For example, if the last step in a synthesis involves precipitating a drug product as a hydrochloride salt, the stoichiometry of the salt formation could be important in establishing control of the process and the identity of product. For a further discussion of these techniques, the reader is referred to the book chapter by Parente [7].

c) Assay: "A specific, stability-indicating procedure should be included to determine the content of the new drug substance. In many cases it is possible to employ the same procedure (e.g., HPLC) for both assay of the new drug substance and quantitation of impurities. In cases where use of a non-specific assay is justified, other supporting analytical procedures should be used to achieve overall specificity. For example, where titration is adopted to assay the drug substance, the combination of the assay and a suitable test for impurities should be used."

With regard to pharmacopeial requirements, the European Pharmacopoeia generally prefers a titration procedure, which is more precise than a chromatographic procedure but less selective, while the USP generally opts for a less precise but more selective, stability-indicating chromatographic procedure.

d) Impurities: Three types of impurities include organic impurities, inorganic impurities, and residual solvents.

   Organic impurity specifications should include specified impurities, unspecified impurities, and total impurities. These impurities may be further characterized as degradation products or process-related impurities or both. For chiral drug substances, the specification should include a test for chiral impurities generally using chiral chromatography, which is preferred, or a test for optical rotation. Specifications for residual solvents usually focus on solvents used in late-stage synthesis that are likely to be present and solvents considered to be restricted due to high toxicity.

   Inorganic impurities are usually controlled by a specification for residue on ignition, and if catalysts or metal-containing reagents are used in the synthesis, a specific test for the metal element should be considered for inclusion in the specifications. At this time, the outmoded testing of heavy metals, conducted by sulfide precipitation, is being phased out in lieu of a more sensitive and selective test typically conducted by Inductively Coupled Plasma-Mass Spectrometry (ICP-MS) or atomic absorption (AA) spectroscopy. For additional information regarding elemental impurities, the reader is referred to USP General Chapter <232>Elemental Impurities – Limits and USP General Chapter <233>Elemental Impurities – Procedures.

6.5.1.2 New Drug Products

The following tests and acceptance criteria are considered generally applicable to all new drug products.

a) Description: "A qualitative description of the dosage form should be provided (e.g., size, shape, and color). If any of these characteristics change during manufacture or storage, this change should be investigated and appropriate action taken. The acceptance criteria should include the final acceptable appearance. If color changes during storage, a quantitative procedure may be appropriate."

   As an example, a typical description for a tablet would be "Pink, round, biconvex, film-coated tablet, with a score on one side and '347' debossed on the other side." As another example, a typical description for a capsule would be "Size 1, gelatin capsule, red cap with '555' black printing, black body, containing a white to off-white powder." The description should be sufficiently detailed to unambiguously identify the drug product. This is particularly important for surveillance of counterfeiting and the identification in the event of potential poisoning or misuse. There are several on-line services that can provide rapid identification of drug products for these purposes.[5] For a capsule, any change in the color of the capsule shell or the capsule fill observed during a stability program or during the shelf-life of the product should be investigated.

b) Identification: "Identification testing should establish the identity of the new drug substance(s) in the new drug product and should be able to discriminate between compounds of closely related structure which are likely to be present. Identity tests should be specific for the new drug substance, e.g., infrared spectroscopy. Identification solely by a single chromatographic retention time, for example, is not regarded as being specific. However, the use of two chromatographic procedures, where the separation is based on different principles, or combination of tests into a single procedure, such as HPLC/UV diode array, HPLC/MS, or GC/MS, is generally acceptable."

   If a spectroscopic method is used for identification, often interferences from excipients in the formulation require a pre-extraction of the active ingredient. In some cases, instead of a complete spectral match as would be required for an IR identification procedure for a drug substance, acceptance criteria involving a selection of five to six significant IR absorbance bands could be specified as indicative of a positive identification.

c) Assay: "A specific, stability-indicating assay to determine strength (content) should be included for all new drug products. In many cases it is possible to employ the same procedure (e.g., HPLC) for both assay of the new drug substance and quantitation of impurities. Results of content uniformity testing for new drug products can be used for quantitation of drug product strength, if the methods used for content uniformity are also appropriate as assays. In cases where use of a non-specific assay is justified, other supporting analytical procedures should be used to achieve overall specificity. For example, where titration is adopted to assay the drug substance for release, the combination of the assay and a suitable test for impurities can be used. A specific procedure should be used when there is evidence of excipient interference with the non-specific assay."

d) Impurities: "Organic and inorganic impurities (degradation products) and residual solvents are included in this category. Refer to the ICH Guidelines Impurities in New Drug Products and Residual Solvents for detailed information. Organic impurities arising from degradation of the new drug substance and impurities that arise during the manufacturing process for the drug product should be monitored in the new drug product. Acceptance limits should be stated for individual specified degradation products, which may include both identified and unidentified degradation products as appropriate and total degradation products. Process impurities from the new drug substance synthesis are normally controlled during drug

5 For example, see the "RxList Pill Identifer" at www.RXList.com.

substance testing, and therefore are not included in the total impurities limit. However, when a synthesis impurity is also a degradation product, its level should be monitored and included in the total degradation product limit. When it has been conclusively demonstrated via appropriate analytical methodology that the drug substance does not degrade in the specific formulation, and under the specific storage conditions proposed in the new drug application, degradation product testing may be reduced or eliminated upon approval by the regulatory authorities."

Specifications for organic impurities should include specified, unspecified, and total degradation products. Process impurities that are not degradation products are not included in the drug product specification since they are controlled in the drug substance specification. If chiral impurities are also degradation products, they should be included in the specification. The drug product should also include specifications for residual solvents and elemental impurities based on the permitted daily exposure limits in ICH Q3C on residual solvents [8] and ICH Q3D on elemental impurities [9], respectively.

6.5.2 Specific Tests

6.5.2.1 Drug Substances

Specific tests commonly included in drug substance specifications are given in Table 6.3. Some of the tests are related to the source of the drug substance or

Table 6.3 Typical specific tests commonly included in drug substance specifications.

Particle size	Tensile strength
Polymorphic forms	Viscosity
Water	Acid-neutralizing capacity
Loss on drying	Color and achromicity
Melting point	Completeness of solution
Specific gravity	Water content
Optical rotation	Thermal analysis
Refractive index	Microbial attributes[a]
pH of solution	Endotoxins[a]
Residue on ignition	Bulk and tap density
Porosity	Flowability[b]
Specific surface area	Dust exposivity index[b]
Enantiomeric purity	

a) Tests for active ingredients used for parenteral administration.
b) Performance tests usually included as in-house specifications.

the route of administration. For example, tests for microbial attributes or endotoxins may be included in the specifications for a drug substance if intended to be used in a drug product for parenteral administration but may not be required if the drug product is for oral administration. To aid in the selection of tests to be included in a comprehensive specification for a drug substance or drug product, the reader is referred to ICH guideline Q6A, which contains a number of useful "decision tree" flow diagrams to aid in the selection of appropriate test.

6.5.2.2 Drug Products

Specific tests commonly included in drug product specifications for selected drug product dosage forms [10] are shown in Table 6.4. As shown in the table, the selection of tests depends on the route of administration.

For additional information on tests and procedures, the reader is referred to the "General Chapters: Chapter Charts" in the USP, which provides an index and cross-references for testing methods [3].

Table 6.4 Specific test for drug products.

Dosage form	Commonly included specific tests	Examples of other specific tests
Oral solids	Dissolution Uniformity of dosage units Residual solvents Elemental impurities	Disintegration Hardness Friability Water content Microbial attributes
Oral solutions and rectal solution	Uniformity of dosage units pH Residual solvents Elemental impurities	Microbial limits Antimicrobial preservative content Alcohol content Specific gravity Deliverable volume
Inhalation	Dose uniformity over the entire contents Residual solvents Elemental impurities	Particle size Alcohol content pH
Injection and for injection	Uniformity of dosage units pH Sterility Bacterial endotoxins Particulate matter Residual solvents Elemental impurities	Minimum fill Water content Antimicrobial effectiveness Antimicrobial preservative content Osmolality Reconstitution time

Table 6.4 (Continued)

Dosage form	Commonly included specific tests	Examples of other specific tests
Topical semisolids	Uniformity of dosage units Residual solvents Elemental impurities	Drug release Minimum fill Microbial limits Alcohol content Particle size distribution Specific gravity
Topical solutions	Uniformity of dosage units pH Antimicrobial preservative content Residual solvents Elemental impurities	Drug release Microbial limits Alcohol content Specific gravity Deliverable volume
Ophthalmic semisolids	Uniformity of dosage units pH Sterility Residual solvents Elemental impurities	Drug release Microbial limits Particle size distribution Specific gravity Minimum fill
Ophthalmic solutions	Uniformity of dosage units pH Sterility Particulate matter Residual solvents Elemental impurities	Microbial limits Dissolution Specific gravity Antimicrobial effectiveness Antimicrobial preservative content Osmolality Deliverable volume
Oral suspension and rectal suspensions	Uniformity of dosage units pH Antimicrobial preservative content Residual solvents Resuspendability Elemental impurities	Drug release Microbial limits Alcohol content Particle size distribution
Suppositories	Uniformity of dosage units Residual solvents Elemental impurities	Drug release Microbial limits
Transdermal systems	Drug release Uniformity of dosage units Residual solvents Elemental impurities	

6.6 Establishing Acceptance Criteria

After the appropriate tests and procedures of analysis have been selected, the next step in developing a specification is to establish meaningful acceptance criteria. Ideally, "the acceptance criteria allow for analytical error, for unavoidable variations in manufacturing and compounding, and deterioration to an extent considered acceptable under practical conditions." [11] They should provide an assurance from a quality perspective that the process remains in a state of control that can be linked back to clinical studies supporting the safety and efficacy of the product. They also need to account for the capability of the manufacturing process and the stability profile of the product over its shelf life. As a prelude to a discussion of establishing release and shelf-life specifications, some basic concepts need to be considered. In particular, the rounding rules applicable to numerical test results and statistical measures to establish attainable and meaningful specifications are discussed. The statistical measures include the confidence interval, the prediction interval, and the tolerance interval. It is also appropriate to acknowledge that meaningful acceptance criteria are often defined by predetermined expectations, industry standards, and regulatory practices to ensure that the product is of adequate quality for its intended use, for example, 90.0–100% of label claim for a typical oral tablet. Thus, evaluation of process capability and the use of statistical measures often confirm that standards for product acceptability can be achieved.

6.6.1 Rounding Rules

For quantitative tests, in order to determine the conformance to specification, the observed result needs to be rounded to the appropriate number of decimal places. For the purpose of determining compliance according to the USP [12]: "The observed or calculated values shall be rounded off to the number of decimal places that is in agreement with the limit expression. Numbers should not be rounded until the final calculations for the reportable value have been completed. Intermediate calculations (e.g., slope for linearity) may be rounded for reporting purposes, but the original (not rounded) value should be used for any additional required calculations. Acceptance criteria are fixed numbers and are not rounded. When rounding is required, consider only one digit in the decimal place to the right of the last place in the limit expression. If this digit is smaller than 5, it is eliminated and the preceding digit is unchanged. If this digit is equal to or greater than 5, it is eliminated and the preceding digit is increased by 1."

Examples of using the rounding rules published in the USP [12] to determine the compliance with the upper and lower limits for an assay and to determine the compliance with an impurity limit test are given in Table 6.5. Note that by USP convention, when the digit of interest is 5, the preceding value is

Table 6.5 Illustration of rounding numerical values for comparison with numerical acceptance criteria.

Compendial requirement	Unrounded value	Rounded result	Conforms
Assay limit ≥98.0%	97.96%	98.0%	Yes
(Assay limit NLT 98.0%)	97.92%	97.9%	No
	97.95%	98.0%	Yes
Assay limit ≤101.5%	101.55%	101.6%	No
(Assay limit NMT 101.5%)	101.46%	101.5%	Yes
	101.45%	101.5%	Yes
Limit test ≤0.02%	0.025%	0.03%	No
(NMT 0.02%)	0.015%	0.02%	Yes
	0.027%	0.03%	No
Limit test ≤3 ppm	3.5 ppm	4 ppm	No
(NMT 3 ppm)	3.4 ppm	3 ppm	Yes
	2.5 ppm	3 ppm	Yes

always rounded up, which introduces a slight statistical bias.[6] However, this policy is consistent with the way most calculators and computers round the value 5.

6.6.2 Statistical Estimation

In establishing acceptance criteria, the total variability of the product should be taken into account. The total variability for any product can have several contributing factors as shown in Eq. (6.1), but the predominant factors that should be considered include the variance in the process and the variance of the measurement.

$$\sigma^2_{\text{Total}} = \sigma^2_{\text{Process}} + \sigma^2_{\text{Measurement}} + \sigma^2_{\text{Ingredients}} + \sigma^2_{\text{Environment}} + \cdots \quad (6.1)$$

where,

σ = the standard deviation of the process or other indicated factor
σ^2 = variance of the process or other indicated factor (variances are additive).

Unless there is knowledge to the contrary, the total variability of the product is assumed to be randomly distributed and it is common to describe the

6 By another convention, the preceding digit would be rounded up if the rounding resulted in an even number and dropped if it resulted in an odd number, which eliminates the statistical bias.

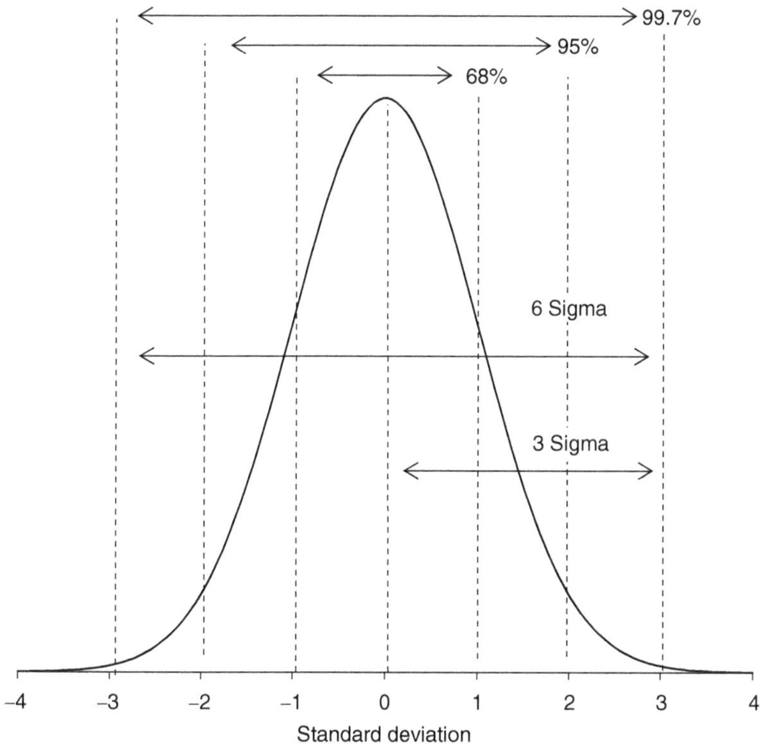

Figure 6.1 Normal (bell-shaped) data distribution curve.

variability using a normal distribution of values about the mean for a given attribute of interest. Figure 6.1 shows the properties for a normal distribution of data about the mean value for a given product attribute. The curve shows the percentage of individual observations that can be expected as a function of the standard deviation from the mean. For example, as shown in the figure, 99.7% of the values would be expected to fall within $\pm 3\sigma$ (i.e., six sigma range) of the mean value. Thus, the standard deviation can be used to establish limits that the attribute would be expected to meet with a stated level of statistical confidence. The suitability of the limits established in this manner will improve as the quantity of attribute data available for calculating the standard deviation increases, preferably from multiple lots of product.

Specific statistical methods are available to separately estimate the process variability and the measurement variability, such as Analysis of Variance (ANOVA) or Gauge R&R. One value in doing this is to understand which source of variation is greatest so that improvement efforts can be directed toward reducing the source contributing the greatest variation.

6.6.2.1 Confidence Interval

A method to estimate the mean of a normal distribution is using the confidence interval [13] of the mean. The confidence interval is the range of values which is likely to include the mean of a given product quality attribute at a specified level of confidence (i.e., 95% confidence). The confidence interval (CI) of the mean, $\bar{x}$, is given by Eq. (6.2):

$$\mathrm{CI} = \bar{x} \pm \frac{ts}{\sqrt{n}} \tag{6.2}$$

where,

$\bar{x}$ = mean value of a given attribute
n = number of observations
t = Student's t value for $(n-1)$ degrees freedom for a given confidence level
s = standard deviation

The confidence interval is important if a specification involves comparing the mean value of a set of observations to the acceptance criteria. The confidence interval for the mean is therefore an appropriate statistic for expressing the result of replicate analyses, but is not useful for establishing acceptance criteria for individual observations.

6.6.2.2 Prediction Interval

Unlike the confidence interval which is the range wherein the average product attribute is most likely to lie, the prediction interval [13] is the range in which the next new observation or measured value is expected with a stated probability equal to the confidence level chosen for the Student's t value. The prediction interval (PI) for the next new observation is given by Eq. (6.3):

$$\mathrm{PI} = \bar{x} \pm ts\sqrt{\left(1 + \frac{1}{n}\right)} \tag{6.3}$$

where,

$\bar{x}$ = mean value of a given attribute
n = number of observations
t = Student's t value for $(n-1)$ degrees freedom for a given confidence level
s = standard deviation

Because of the added uncertainty of predicting a new, single value, the prediction interval is always wider than the confidence interval. The prediction interval only bounds a single future value, whereas a tolerance interval, discussed in the next section, bounds the entire population of future values [14].

6.6.2.3 Tolerance Interval

The *statistical*[7] tolerance interval is intended to bound multiple future values, that is, values for future product production, making it the appropriate choice of the intervals discussed in this chapter for establishing acceptance criteria. The tolerance interval is the range of values that is likely to include a given portion of a specific attribute, such as an assay value, with a given probability, for example, the range of values that 95% of the observations will fall within with 95% confidence interval. Unlike the confidence interval, which is a predictor for the mean value, the tolerance interval is a useful tool for predicting the conformance of individual values to an acceptance criterion.

Using this approach for a normal distribution, the two-sided, tolerance interval (TI) limits are given by the equation:

$$\mathrm{TI} = \bar{x} \pm ks \qquad (6.4)$$

where,

$\bar{x}$ = sample mean

s = standard deviation estimating the true standard deviation of the population, σ

k = tolerance factor for the confidence level required and the percent of observations in the coverage

The tolerance factor, k, which can be derived from the z-score,[8] is related to the number of standard deviations needed to include the population for a given percentage of observations with a given confidence level based on a given number of observations. The calculation of the factor, k, is complex and will not be addressed here, but reference tables of values [15] are available and calculation of tolerance limits are included in many commercial statistical analysis software packages[9] and found on the Internet.[10]

An important difference between the confidence interval and the tolerance interval is that the confidence interval is an estimate of the mean value of an attribute, while the tolerance interval, similarly to the prediction interval, estimates the individual values. Most often, testing for the purpose of evaluating the conformance is performed as singlet or duplicate determinations, and the

7 A *statistical* tolerance limit is calculated from process information as opposed to an *engineering* tolerance limit that is used to establish operating ranges based on an engineering design.

8 The z-score is the number of standard deviations from the mean required to include a given percentage of a normal distribution. For example, $\mu \pm 1.96\sigma$ would include 95% of the area for a normal distribution. The z-score differs from the tolerance factor in that it does not include a confidence level estimation.

9 One such package is Minitab® 16 Statistical Software available from Minitab Inc., State College, PA. URL: www.minitab.com.

10 For example, see http://statpages.info/tolintvl.html.

acceptance criteria are most appropriately determined using a tolerance interval approach (see Eq. (6.4)).

As stated previously, statistical estimates improve as the amount of data they are based on increases, that is, as the degrees of freedom increase. Germane to statistical intervals is the knowledge that as the degrees of freedom approach infinity, the prediction and tolerance intervals become equal [16]. One issue in using any statistical approach is that a sufficient amount of data needs to be available to get valid estimates. However, usually only limited data are available at the time of filing a marketing authorization. In some cases, it may be advantageous to establish an interim specification until additional data become available to more accurately access the process capability.

6.6.2.4 Monte Carlo Simulation of Quality Attributes

As the name of the famous casino implies, Monte Carlo simulation is a probabilistic technique for estimating the variability of a given quality characteristic. One drawback of using statistical methods to develop acceptance criteria is that sufficient data need to be available to apply the techniques. Often, at the time of filing a regulatory submission, only limited data regarding the process capability are available. Using Monte Carlo techniques, one can generate a large data set of outcomes to simulate the expected variability resulting from the interaction of critical variables contributing to the overall attribute variability. To use the technique, one has to (1) identify the critical variables and (2) develop a mathematical model to define the contribution of the expected variability of each critical variable to the overall variability of the quality characteristic. Next, the probable overall variability of a given attribute is calculated from the probable range of variability for each variable by randomly generating numbers for a given distribution believed to be associated with the probability of occurrence (e.g., normal distribution) about the variable mean value. In this way, thousands of outcomes can be simulated that can be used to estimate the overall variability of a quality attribute. In some cases, the models relating the variables can be derived from design of experiment (DOE) studies. To cite a number of examples of application, the use of Monte Carlo methods has been applied to risk analysis in pharmaceutical product design [17], determination of in-process limits for parenteral solution manufacturing [18], evaluation of measurement uncertainty of pharmaceutical certified reference materials [19], and establishing dissolution specifications [20], and other applications [21] [22]. For a good general reference of the use of Monte Carlo simulation methods in the pharmaceutical industry, the reader is referred to the book by Chang [23].

6.6.3 Establishing Acceptance Criteria Limits

Approaches for establishing acceptance criteria can broadly be separated into two general cases. The first case involves setting acceptance criteria for an attribute that does not change with time for a given batch of product. The second case considers attributes that change with time for a given batch of product due to storage and environmental conditions. Understanding time-related change involves trend analysis to determine product shelf life. Following is a discussion of both approaches.

6.6.3.1 Acceptance Criteria for Attributes that Do Not Change with Time

For attributes that do not change with time, the acceptance criteria are determined using the combined uncertainty of all the contributions to the process and measurement variability. Based on the recommendation in ICH guideline Q6A, limits are commonly derived using the so-called 3σ approach in which the lower specification limit (LSL) and the upper specification limit (USL) of the acceptance range are based on a variation of ±3 standard deviations for a given quality attribute (see Eq. (6.5)) corresponding to 99.7% coverage for a normal distribution of data (see Figure 6.1), which are similar to a tolerance interval where $k = 3$ (i.e., 50% confidence/99.7% coverage) in Eq. (6.4)

$$\left(\text{LSL},\text{USL}\right) = \bar{x} \pm 3s \quad (6.5)$$

A comparison of using different statistical approaches to setting acceptance criteria based on the data for a given attribute is shown in Table 6.6. If the

Table 6.6 Comparison of statistical estimation methods.

Data set: 4.02, 3.81, 3.92, 4.23, 4.31, 3.99, 3.82, 4.42, 4.23
Mean = 4.083
Standard deviation = 0.2211

Statistic	Acceptance criteria	Lower limit	Upper limit
Confidence interval (95% confidence)	4.08 ± 0.14	3.94	4.22
Confidence interval (99% confidence)	4.08 ± 0.19	3.89	4.27
Prediction interval (95% confidence)	4.08 ± 0.39	3.69	4.47
Tolerance interval 90% confidence level 95% observations covered	4.08 ± 0.69	3.39	4.77
Tolerance interval 95% confidence level 95% observations covered	4.08 ± 0.78	3.30	4.86
Tolerance interval 99% confidence level 95% observations covered	4.08 ± 1.01	3.08	5.09
Tolerance interval "3σ estimator"	4.08 ± 0.66	3.42	4.74

confidence interval of the mean value is used, even at the 99% confidence level, which gives wider limits compared to the 95% level, the number of individual observations still falls outside the upper and lower limits. On the other hand, the limits derived from the tolerance interval estimates better reflect the capability of the process for individual units. In addition, note that the "3σ" estimate," which is based on ±3 standard deviations about the mean, gives estimates similarly to the tolerance interval for $k = 3$ ($n = 9$). Thus, for cases were compliance is measured based on individual observations, the tolerance limit gives a more realistic estimate of acceptance criteria limit by incorporating the statistical variation of the individual values.

6.6.3.2 Acceptance Criteria for Attributes that Change with Time–Trend Analysis

For attributes that change with time, in addition to process and measurement variability, the acceptance criteria need to include the rate of change of the attribute and an estimate of the corresponding uncertainty. As previously stated, the shelf life is defined as the time period over which the attribute will meet the acceptance criteria with 95% confidence interval. Figure 6.2 shows a trend analysis of the change in assay (i.e., % label claim) with time for the first

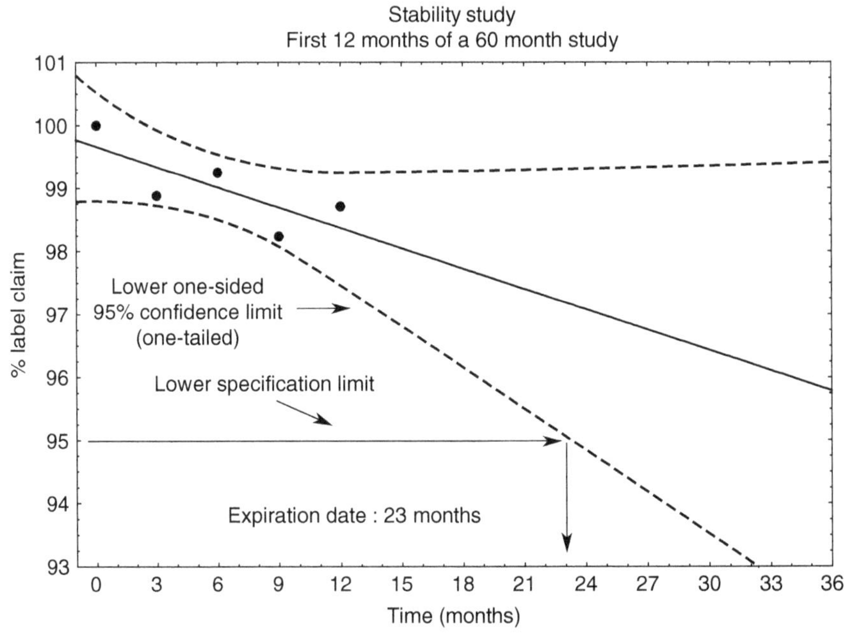

Figure 6.2 Stability trend line for the first 12 months of a 60-month stability program.

12 months of a 60-month stability program.[11] The points were selected to represent typical stability sampling intervals. Shown in the figure is the trend line for a linear regression model with the two-sided, 90% confidence bands limits for the model. Note that since the assay decreases with time, the lower confidence band represents the one-sided, 95% confidence limit. The expiration date is determined where the LSL (i.e., 95% of label claim) intersects the lower confidence bound at approximately 23 months. Note that as the data are extrapolated past the last observed value, the confidence bands rapidly diverge from the trend line, giving rise to significantly shorter estimates of the shelf life than the trend line.

Figure 6.3 shows the trend line estimated for data through 60 months of the stability program. As shown in the figure, as more data become available, using the 95% one-sided confidence band for the regression line, a shelf life of 42 months is estimated. Thus, while the slope of the regression trend line has not changed significantly, the precision in the estimate of the fitted model is significantly improved. The result of this improved model is that shelf-life estimate is extended. Thus, shelf-life estimates based on limited data often underestimate the long-term shelf life allowing an opportunity to extend the shelf life

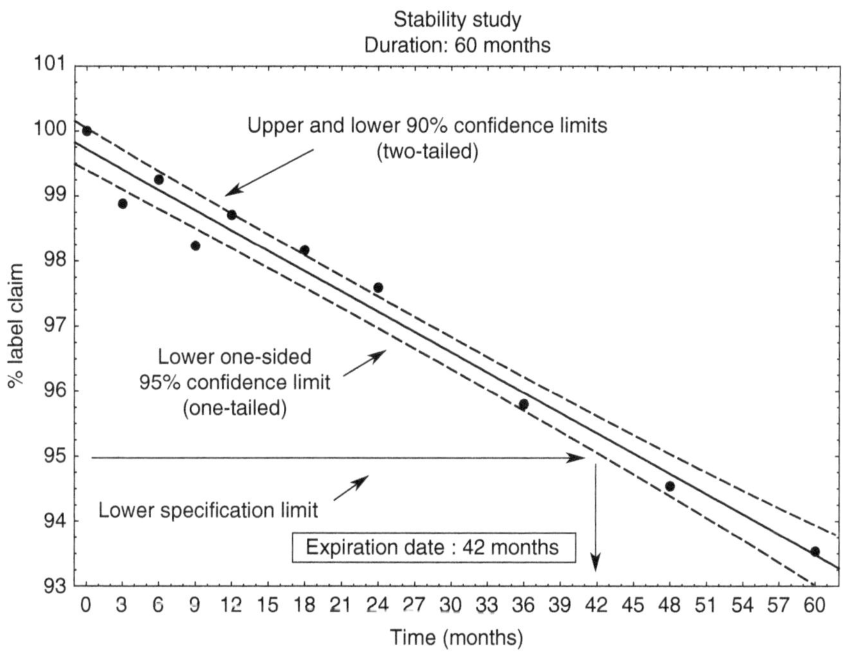

Figure 6.3 Stability trend line for a 60-month stability program.

11 For the stability program, data at 0, 3, 6, 9, and 12 months were used.

as more long-term data become available. This is important in filing applications for new drug products since only 12 months of data are typically available at the time of filing. As more data become available, there is a potential to increase the product shelf life.

6.7 Release Specifications

Release specifications are generally derived from process capability information. In this section, approaches for establishing release specification for a given attribute based on process capability considerations are discussed. The goal is to establish acceptance criteria for a given attribute that will consistently ensure that this attribute, for example, assay, will meet the acceptance criteria with a predictable probability. For attributes that do not change with time, similar techniques can be used for establishing shelf-life specifications. For attributes that increase or decrease with time, the stability profile and the expected expiration date will need to be taken into consideration to establish an upper or lower limit at release.

Figure 6.4 depicts three distributions for a given process, one with low variability, one with medium variability, and one with high variability about a mean

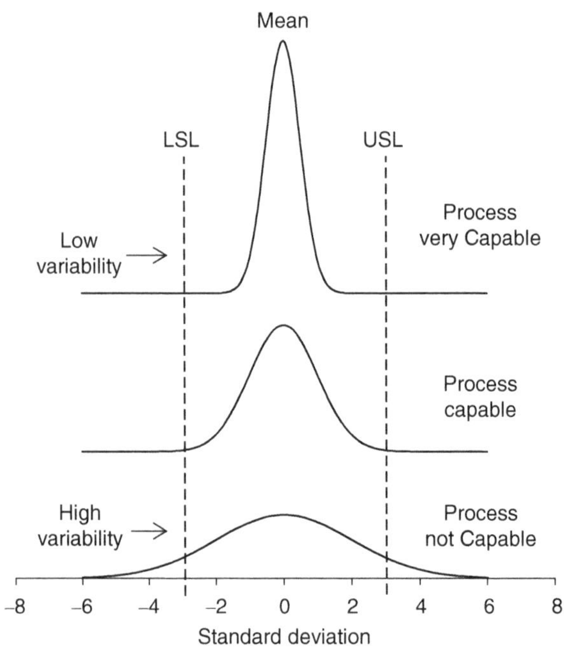

Figure 6.4 Process capability.

value for an attribute that can be modeled using a normal distribution. In addition, shown in the figure are the LSL and the USL for the attribute. Since in the case with low variability, the distribution falls within the LSL and USL values, the process is considered very capable. Conversely, the process with high variability would not be considered very capable of consistently producing a material that met the specification limits. It should be noted that the process capability is related to the specification limits. If the specification limits were tightened, a process that was capable could become not capable while if the limits were widened, a process that was not capable could become capable of meeting the specifications. In addition, for the process considered "capable" at the $\pm 3\sigma$ level, it could become "not capable" if the process mean shifted. Thus, additional considerations would need to be included if process or measurement drift was an issue.

6.7.1 Using the Process Capability Index to Estimate Attribute Acceptance Criteria

Assuming a variation of ±3 standard deviations from the mean, a process capability index, Cp [24], can be defined as

$$\mathrm{Cp} = \frac{\mathrm{USL}-\mathrm{LSL}}{6\sigma} \tag{6.6}$$

where,

USL = upper specification limit
LSL = lower specification limit
σ = standard deviation

While it is somewhat arbitrary, if a value of $\mathrm{Cp} > 1.3$ indicates that the process is capable of achieving the specification limits and if the standard deviation is known, one can calculate the difference between the USL and the LSL as

$$(\mathrm{USL}-\mathrm{LSL}) = \mathrm{Cp}\times 6\sigma = 7.8\sigma \tag{6.7}$$

If one assumes that the limits are symmetrical about the mean value for a given attribute, the limits can be estimated as

$$(\mathrm{LSL},\mathrm{USL}) = \bar{x} \pm \frac{7.8\sigma}{2} = \bar{x} \pm 3.9\sigma \tag{6.8}$$

For example, assuming a symmetrical distribution about the mean, if an assay value for a tablet has a mean of 100.0% label claim and a standard deviation of 1.28, using Eq. (6.8),

$x = 100.0 \pm 4.99$, which supports setting an acceptance criteria of 95.0–105.0% of label claim.

There are a number of commercially available statistical programs that automate the calculation of the process capability and provide USL and LSL.[12]

6.8 Relationship between Release and Shelf-Life Specifications

While release specifications are derived from process capability, a number of other factors need to be taken into consideration to develop a meaningful shelf-life specification. A number of authors have addressed this complex problem [25] [26] [27] [28]. Three important factors are as follows: (1) the uncertainty in the estimate of the attribute of interest, for example, assay, impurities; (2) the estimate of the change of the attribute with time; and (3) the estimate of the uncertainty in the change in the attribute with time. With regard to the estimate of attribute uncertainty, often this includes the batch-to-batch variation in the attribute plus the associated measurement uncertainty. The contribution of these factors in determining the shelf-life and release specifications for an assay determination that shows a loss over time is shown in Figure 6.5. Thus, if the lower release limit (LRL) and the upper release limit (URL) can be estimated from the process capability, the lower and upper stability limits can be determined with the knowledge of the loss with time, an estimate of the uncertainty in the loss and the uncertainty in batch and assay estimate. For attributes that do not change with time, the estimate of the loss and the uncertainty in the loss can be neglected, and the upper and lower stability limits can be estimated using the process capability and the uncertainty in the batch and the assay.

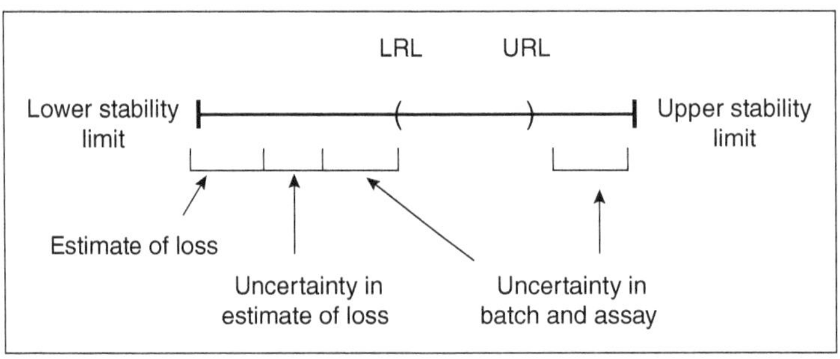

Figure 6.5 Relationship between release and shelf-life specifications for an assay determination. *Source*: Reproduced with kind permission of Laura [28].

12 One such package is Minitab® 16 Statistical Software available from Minitab Inc., State College, PA. URL: www.minitab.com.

Practically, the ability to predict these parameters is affected by the limited data that are available at the time of filing a drug authorization application. However, for some tests, it may be possible to achieve marketing authorization with limited data reflecting a somewhat worst-case analysis pending revision when additional full-scale process data become available. An additional complication is that data obtained from multiple stability studies could show different rates of loss so that a worst-case estimate would need to be used. The FDA has also provided recommendations on how to evaluate stability data based on recommendations in ICH guideline Q1E [29], which should be considered in developing and justifying specification acceptance criteria.

The graphical representation of the model in Figure 6.5 is given in Figure 6.6 for an assay that decreases with time. As shown in the figure, the batch and assay uncertainty is estimated from the initial data or process capability, the estimate of the loss is determined from the linear regression trend line, and the uncertainty in the estimate of the loss is determined from the lower one-sided, 95% confidence band.

The example given in Figure 6.6 is derived from the data for only one batch. Typically, data from multiple batches and packaging configurations need to be considered since they may have different slopes and initial values. For example, if the samples tested had similar slopes for different lots but different initial values, a "worst-case" estimate could be developed by constructing a line with a parallel slope to the one shown in the figure but starting at the lower

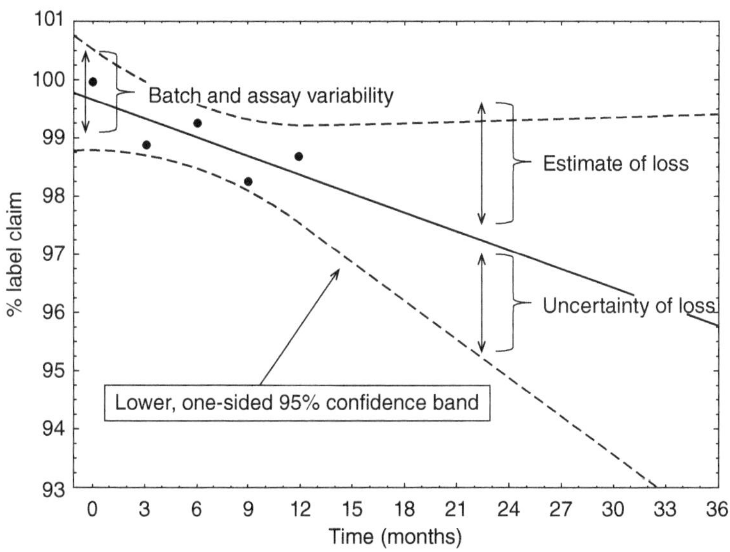

Figure 6.6 Graphical representation of the uncertainties to consider between shelf life and release for an assay determination.

limit of the batch and assay variability estimate. For methods for determining if the slope or initial value can be pooled, the reader is referred to ICH guideline Q1E [21].

6.9 Using a Control Chart for Trend Analysis

After acceptance criteria have been established for a given quality attribute, monitoring the conformance of the process to specifications over time can be done using control charting techniques commonly used for statistical process control [30]. While there are many control charting approaches, one of the simplest is to plot the attribute with time and monitor the data trend relative to historical or predetermined limits. Shown in Figure 6.7 is a control chart for the trend of assay data for the percent label claim of a drug product with time or consecutive lot numbers of production. The chart also shows lines giving the specification limits, an alert limit, and an action limit. The alert limit and the action limits are determined using the standard deviation, σ, of the process. In this case, the alert limit is set at a value of $\pm 2\sigma$ and the action limit is set at $\pm 3\sigma$. However, other limits can be established as appropriate based on a risk assessment of the stability of the process. The goal is to maintain the process in a state of control and take action when there is evidence of process change that left unchecked could result in batch failures. Typically, results found above the

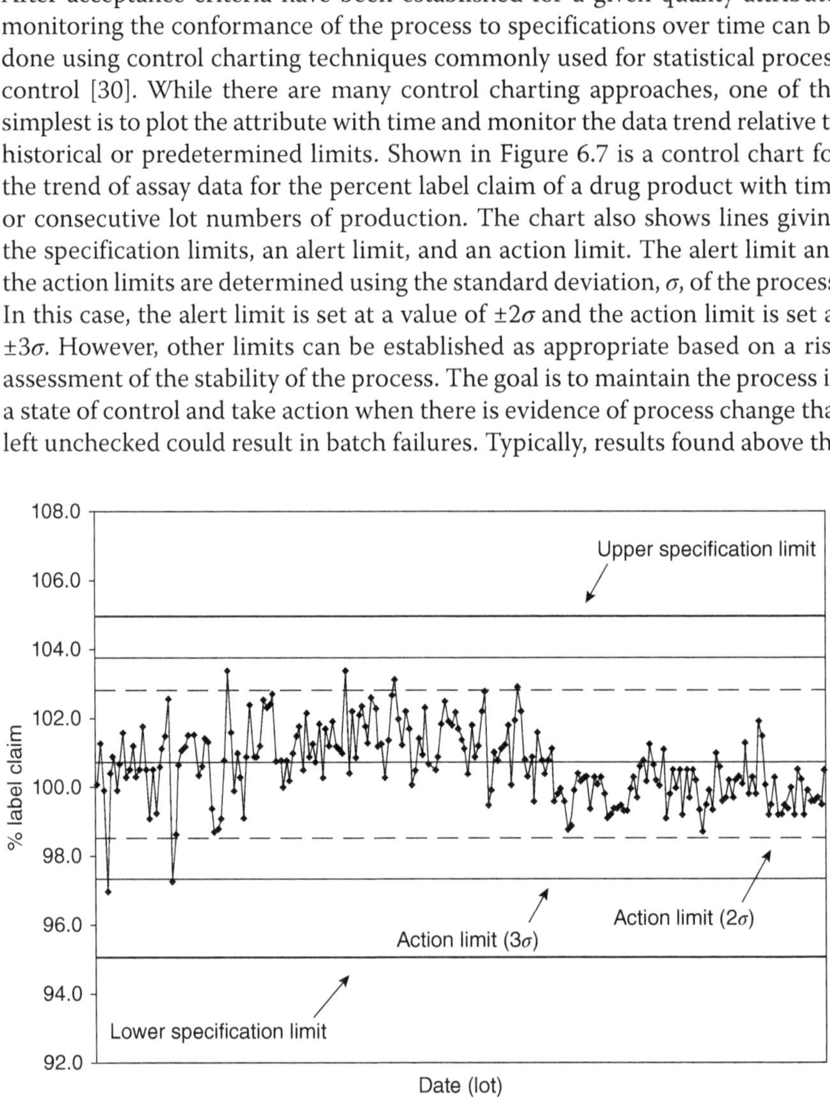

Figure 6.7 Control chart.

alert limit are used as early warning signals that either the process or the analytical testing may be trending out of control. A failure of the action limit usually involves a more rigorous investigation with further preventive actions taken to bring the process back into control. In addition, note in the example that since drug products are typically formulated at 100% of label claim, while there may be occasional deviations, the results should average 100% over time, and the failure to meet this should be investigated.

6.10 Life Cycle Management of Specifications

6.10.1 Approach to Life Cycle Management

Once meaningful specifications have been established, there is a regulatory expectation that tests and analytical procedures are periodically reviewed to establish fitness for use. This may involve periodic revalidation, trending of results, reviewing the impact of changes or improvements to the procedure, and reviewing any failures that may be due to the analytical method. The FDA has published a "Guidance for Industry" detailing expectations for the life cycle management of analytical procedures [31].

6.10.2 Impact of the Investigation of Out-Of-Specification (OOS) and Out-Of-Trend (OOT) Results on Test Methods and Specifications

As part of the analytical procedure life cycle management, it is important to review the relevance of procedures and specifications in the event of OOS/OOT findings. The FDA has issued a guidance for industry [32] delineating the FDA's expectations to conduct an investigation of aberrant or suspect results with the objective of discovering the underlying root cause of the results, that is, lab-related or manufacturing-related, so that the appropriate corrective and preventive actions (CAPAs) can be instituted to prevent recurrence or to remove adulterated/misbranded products from moving in commerce. As of this writing, the failure to conduct meaningful investigations is still one of the major sources of regulatory audit observations.. Thus, a strong OOS/OOT quality system is an important part of product life cycle management that can detect changes that may have occurred in either the analytical test or manufacturing of drug substances and drug products.

As mentioned earlier, the root cause of the suspect result can generally be related to either lab-related or manufacturing-related issues. To determine the root cause of the OOS/OOT result, the investigation takes place in two phases: the lab phase and the manufacturing phase. If the root cause is found to be related to laboratory testing, remedial and CAPA should be taken. While it is beyond the scope of this chapter to discuss OOS/OOT investigations in detail, a brief description of the process is provided as follows.

Level 1 – Lab-Phase Investigation
The Level 1 investigation takes place is two stages. In the first stage of the investigation, results are reviewed with the intent of identifying obvious errors or assignable causes. This phase of the investigation is usually conducted using a checklist format and documented on a preliminary investigation form. If no obvious causes are identified, a second stage involving a more in-depth lab investigation is then conducted to verify the initial observation that was considered to be aberrant. If the initial result cannot be verified, the most likely root cause is identified, which may involve a re-evaluation of the method's continuing fitness for use. If the result is verified, the lab phase of the investigation would be concluded, and the manufacturing stage of the investigation would commence.

Level 2 – Manufacturing-Phase Investigation
The preliminary manufacturing investigation begins at the end of the lab-phase preliminary investigation, and the investigations proceed in parallel since if the root cause is manufacturing-related, the investigation would need to move forward, with time being of the essence to prevent additional batch failures. At this stage, the batch records are reviewed for deviations that could have impacted the product. As part of the investigation, raw materials would also be reviewed. If the root cause of the OOS/OOT is traced to a raw material, additional specifications or modification of existing specifications may be needed to bring the process back into a state of control. Additional in-process controls may also be necessary. This is usually a consequence of the fact that only limited data may be available at the time of filing and not all of the manufacturing variables were identified in the original process validation.

6.11 Summary

The process of establishing specifications for a drug substance or drug product involves selecting a list of tests, appropriate test procedures, and establishing meaningful acceptance criteria that ensure that a substance or drug product meets the standards of identity, strength, quality, and purity throughout its shelf life. The conformance of quality attributes to appropriate specifications ensures that the drug substances, excipients, and drug products are suitable for their intended use. In this chapter, a two-step approach to establishing specifications was given: (1) identifying appropriate tests and suitable analytical test methods and (2) evaluating and justifying acceptance criteria based on process capability and stability characteristics. Finally, the process does not end with the approval of the specification but is dynamic. The life cycle management of product conformance to specifications involves periodic review and trending of data and investigation of the results with the aim of maintaining a state of control, ensuring that high standards of quality are met throughout the life of the product.

Acknowledgments

The author thanks Dr David A. Fay and Mr Myron Diener for providing helpful and thoughtful comments and suggestions on the monograph.

References

1 ICH Harmonized Tripartite Guideline *Specifications: Test Procedures and Acceptance Criteria for New Drug Substances and New Drug Products: Chemical Substances, Q6A*. International Conference on Harmonization of Technical Requirements for Registration for Human Use, Current Step 4 version dated October 6, 1999.

2 ICH Harmonized Tripartite Guideline *Specifications: Test Procedures and Acceptance Criteria for Biotechnological/Biological Products.* International Conference on Harmonization of Technical Requirements for Registration for Human Use, Current Step 4 version dated March 10, 1999.

3 United States Pharmacopeia, USP 39 – NF 34, vol. 1, Rockville, MD: USP Pharmacopeial Convention, 2016.

4 C. M. Riley and L. E. Little, Specifications of Drug Substances and Products – Development and Validation of Analytical Methods, Amsterdam: Elsevier, 2014, pp. 12–15.

5 Food and Drug Administration, Guidance for Industry: PAT – A Framework for Innovative Pharmaceuticl Developmement, Manufacturing and Quality Assurance, Rockville, MD: U.S. Department of Health and Human Services, 2004.

6 General Chapter <1061> Color – Instrumental Methods, in USP 39 – NF 34, vol. 1, Rockville, MD: United States Pharmacopeial Convention, 2016.

7 E. Parente, Description and Identification, in *Specificaitons of Drug Substances and Products – Developmemt and Validation of Analytical Methods*, 1st ed., C. M. Riley, T. W. Rosanske, and S. R. Rabel Riley Amsterdam: Elsevier, 2014, pp. 98–105.

8 International Conference on Harmonization (ICH), "Impurities: Guideline for Residual Solvents Q3C(R5)," February 2011.

9 International Conference on Harmonization (ICH), "Guideline for Elemental Impurities Q3D," Step 4, 16 December 2014.

10 USP Guideline for Submitting Requests for Revisions to USP-NF v3.1, Rockville, MD: United States Pharmacopeial Convention, April 2007.

11 "General Notice 4.10.20. Acceptance Criteria," in *USP 39 – NF 34*, vol. 1, Rockville, MD: United States Pharmocopeial Convention, 2016.

12 "General Notice 7.2. Rounding Rules," in *USP 39 – NF 34*, vol. 1, Rockville, MD: United States Pharmacopeial Convention, 2016.

13 E. J. Dudewicz, Basic Statistical Methods, in *Juran's Quality Handbook*, 5th ed., J. M. Juran and A. B. Godfrey, Eds., New York: McGraw-Hill, 1998, pp. 42–47.

14 K. Krishnamoorthy and T. Mathew, Statisitcal Tolerance Regions: Theory, Applications and Computations, Hoboken, NJ: John Wiley and Sons, Inc., 2009, pp. 1–5.

15 International Organization for Standardization, "Statistical Interpretation of Data – Part 6: Determination of Statistical Tolerance Intervals, ISO 16269-6," 2014.

16 De Gryze S., Langhans I., Vandebroek M. Using the correct intervals for prediction: a tutorial on tolerance intervals for ordinary least-squares regression. Chemom Intell Lab Syst 87(2), [2007]; 147.

17 B. Gural, F. Stanfield and D. Rufino, "Monte Carlo Simulation for Risk Analysis in Pharmaceutical Product Design," in *Proceedings of the 2007 Crystal Ball User Conference*, Denver, 2007.

18 Kuu W. Y., Chilamkurti R. Determination of in-process limits during parenteral solution manufacturing using Monte Carlo Simulation. J Pharm Sci Technol 57(4), [Jul–Aug 2003]; 263–276.

19 Rocha W. F., Nogueira R.. Monte Carlo simulation for the evaluation of measurement uncertainty of pharmaceutical certified reference materials. J Braz Chem Soc 23(3), [2012]; 385–391.

20 Wojcik R. C. Setting dissolution specifications. Dissolution Technol 4(3), [August 1997]; 12–20.

21 Burgess C. Measurement uncertainty without the math. Pharm Technol 40(2), [2016]; 26–40.

22 H. Yang, Use of Monte Carlo Simulation to Address CMC Issues, Institue of Validation Technology, 2013.

23 M. Chang, Monte Carlo Simulation for the Pharmaceutical Industry – Concepts, Algorithms and Case Studies, Boca Raton, FL: CRC Press, Taylor and Francis Group, 2010.

24 F. M. Gyrna, Operations, in *Juran's Quality Handbook*, 5th ed., J. M. Juran and A. B. Godfrey, Eds., New York: McGraw-Hill, 1998, pp. 16–18.

25 Murphy J. R., Hofer J. D. Establishing shelf life, expiry limitss and release limits. Drug Inf J 36, [2002]; 769–781.

26 Allen P. V., Dukes G. R., Gerger M. Determination of release limits: a general methodology. Pharm Res 8(9), [1991]; 1210–1213.

27 J. R. Murphy and D. Weisman, "Using Random Slopes for Estimating Shelf Life," in *Proceedings of the Biopharmaceutical Section Annual Meeting, American Statistical Association*, Anaheim, California, 1990.

28 J. A. Grimes and L. B. Foust, "Establishing Release Limits with a Random Slopes Model," in *Biopharmaceutical Proceedings, American Statistical Association*, Alexandria VA, 1994.

29 Food and Drug Administration, Guidance for Industry – Q1E Evaluation of Stability Data, Rockville, MD: U.S. Department of Health and Human Services, 2004.

30 H. M. Wadsworth, Statistical Process Control, in *Juran's Quality Handbook*, 5th ed., J. M. Juran and A. B. Godfrey, Eds., New York: McGraw-Hill, 1998.

31 Food and Drug Administration, Guidance for Industry – Analytical Procedures and Methods Valication for Drugs and Biologicals, Rockville, MD: U.S. Department of Health and Human Services, 2015.

32 Food and Drug Administration, Guidance for Industry – Investigating Out-of-Specification (OOS) Test Results for Pharmaceutical Production, Rockville, MD: U.S. Department of Health and Human Services, 2006.

7

Impurities

Behnam Davani

7.1 Scope

The impurities in drug substances and drug products must be evaluated and controlled for product safety and quality as a critical part of regulatory requirements. The International Conference on Harmonisation (ICH) has established several guidelines [1–3] on the specification and qualification of impurities in support of the registration applications to regulatory bodies in the United States, European Union, and Japan. This chapter includes topics on definitions, classifications, and limits of impurities based primarily on ICH guidelines.

The procedures for the determination of different types of impurities are regularly updated, and the acceptable limits are adjusted based on new findings concerning the safety and toxicity of the impurities. In addition, new and advanced analytical techniques with more sensitivity and accuracy are developed to detect lower impurity limits. Impurities are not desirable even if they are not toxic. As advanced analytical instruments become more readily available and affordable, they will be used more frequently in QC laboratories. Due to evolving changes in all aspects of impurities (characterization, sources, limits, etc.), it is recommended to keep abreast of the most recent developments in regulatory, compendial, and related guidelines to meet the required specifications.

The focus of this chapter is on impurities in smaller-molecular-weight compounds. Therefore, the impurities in biological/biotechnological products and other related topics such as microbiological contamination and cleaning validation are not covered.

7.2 Definitions

The definitions of organic impurities in drug substance and drug products are included as follows [1, 2]:

Pharmaceutical Analysis for Small Molecules, First Edition. Edited by Behnam Davani.

Identified Impurity: An impurity for which a structural characterization has been achieved.

Identification Threshold: A limit above (>) which an impurity should be identified.

Qualification: The process of acquiring and evaluating data that establishes the biological safety of an individual impurity or a given impurity profile at the level(s) specified.

Qualification Threshold: A limit above (>) which an impurity should be qualified.

Reporting Threshold: A limit above (>) which an impurity should be reported.

Specified Impurity: An impurity that is individually listed and limited with specific acceptance criteria.

Unspecified Impurity: An impurity that is limited by general acceptance criteria but not individually listed with its own specific acceptance criteria.

Unidentified Impurity: An impurity for which a structural characterization has not been achieved.

Degradation Product: An impurity resulting from a chemical change in the drug substance brought about during manufacture and/or storage of the new drug product by the effect of, for example, light, temperature, pH, water or by reaction with an excipient and/or the immediate container closure system.

7.3 Classification of Impurities

The types of impurities include the following [1–3]:

- Organic impurities (process- and drug-related)
- Inorganic impurities
- Residual solvents

Organic impurities can result from the manufacturing process and/or storage and can arise from multiple sources such as starting material, degradation products, reagents, and catalysts. These impurities can be identified (known chemical structure) or unidentified (unknown chemical structure).

Inorganic impurities can arise from the manufacturing process and are usually identified. Some examples of such impurities are reagents, catalysts, inorganic salts, and heavy metals/other elemental impurities.

Residual solvents in pharmaceuticals are organic volatile impurities that are used or produced in the manufacturing of drug substance, excipient, or product. The use of these solvents needs to be controlled, and appropriate limits based on their toxicity and safety must be established. ICH has provided the following classification of residual solvents based on risk assessment to human health [3]:

Class 1 Solvents: These solvents such as benzene, carbon tetrachloride, and other specific chlorinated solvents are known or strongly suspected to cause unacceptable toxicities. The use of such solvents should be avoided.

Class 2 Solvents: Solvents with less toxicity levels such as acetonitrile, toluene, and chloroform. The use of such solvents should be limited.
Class 3 Solvents: Solvents with low toxic potential. These solvents have permitted daily exposure (PDE) of 50 mg or more per day.

The complete list of Class 1, Class 2, and Class 3 solvents with appropriate limits can be found in Refs [3, 4].

7.4 Qualification of Impurities

The qualification and the acceptance criteria for impurities are based on safety consideration. Because impurities are undesirable and do not add any value, the impurity levels may be set at lower than safe limits based on the manufacturing and/or analytical capability of detecting such impurities. The ICH provides the following impurity thresholds in drug substances as highlighted in Table 7.1.

The example in Table 7.2 illustrates the required action to be taken based on the maximum daily dose of 1 g. The corresponding thresholds for this daily dose are as follows:

$$\text{Reporting Threshold} = 0.05\%$$

$$\text{Identification Threshold} = 0.10\% \text{ or } 1.0 \text{ mg per day intake (whichever is lower)}$$

$$\text{Qualification Threshold} = 0.15\% \text{ or } 1.0 \text{ mg per day intake (whichever is lower)}$$

The corresponding reporting, identification, and qualification threshold values for impurities and degradation products in drug products are highlighted in Tables 7.3–7.5, respectively [2].

Table 7.1 Thresholds for impurities in drug substances [1].

Maximum daily dose (g/day)[a]	Reporting threshold (%)[b,c]	Identification threshold[c]	Qualification threshold[c]
≤2	0.05	0.10% or 1.0 mg/day intake (whichever is lower)	0.15% or 1.0 mg/day intake (whichever is lower)
>2	0.03	0.05%	0.05%

a) The amount of drug substance administered per day.
b) Higher reporting thresholds should be scientifically justified.
c) Lower thresholds can be appropriate if the impurity is unusually toxic.

Table 7.2 Example of reporting, identification, qualification of impurities.

Raw data result (%)	Reported result	Calculated total daily intake (TDI) mg of the impurity	Action identification	Action qualification
0.0320	Not reported	0.1	None	None
0.0972	0.10	0.5	None	None
0.1391	0.14	0.7	Yes	None
0.1824	0.18	0.9	Yes	Yes

Table 7.3 Reporting thresholds for impurities and degradation products in drug products.

Maximum daily dose (g)[a]	Reporting threshold (%)[b,c]
≤1	0.1
>1	0.05

a) The amount of drug substance administered per day.
b) Thresholds for degradation products are expressed either as a percentage of the drug substance or as total daily intake (TDI) of the degradation product. Lower thresholds can be appropriate if the degradation product is unusually toxic.
c) Higher thresholds should be scientifically justified.

Table 7.4 Identification thresholds for impurities and degradation products in drug products.

Maximum daily dose[a]	Identification threshold[b, c]
<1 mg	1.0% or 5 μg TDI, whichever is lower
1–10 mg	0.5% or 20 μg TDI, whichever is lower
>10 mg–2 g	0.2% or 2 mg TDI, whichever is lower
>2 g	0.10%

a) The amount of drug substance administered per day.
b) Thresholds for degradation products are expressed either as a percentage of the drug substance or total daily intake (TDI) of the degradation product. Lower thresholds can be appropriate if the degradation product is unusually toxic.
c) Higher thresholds should be scientifically justified.

Table 7.5 Qualification thresholds for impurities and degradation products in drug products.

Maximum daily dose[a]	Qualification threshold[b,c]
<10 mg	1.0% or 50 μg TDI, whichever is lower
10–100 mg	0.5% or 200 μg TDI, whichever is lower
>100 mg–2 g	0.2% or 3 mg TDI, whichever is lower
>2 g	0.15%

a) The amount of drug substance administered per day.
b) Thresholds for degradation products are expressed either as a percentage of the drug substance or as total daily intake (TDI) of the degradation product. Lower thresholds can be appropriate if the degradation product is unusually toxic.
c) Higher thresholds should be scientifically justified.

The example in Table 7.6 illustrates the required action to be taken based on the maximum daily dose of 1.8 mg. The corresponding thresholds for this daily dose are as follows:

Reporting Threshold = 0.05%
Identification Threshold = 0.2% or 2 mg TDI, whichever is lower
Qualification Threshold = 0.2% or 3 mg TDI, whichever is lower

The process impurities are controlled in the drug substance and thus are not usually monitored in the corresponding drug product. These impurities are not expected to increase in the drug products. Therefore, only degradation products and those impurities generated in the final product (e.g., impurities due to interaction between active and excipient, etc.) are controlled.

Table 7.6 Example of reporting, identification, qualification for impurities.

Raw data result (%)	Reported result	Calculated total daily intake (TDI) mg of impurity	Action identification	Action qualification
0.0420	Not reported	1	None	None
0.072	0.07	2	None	None
0.1383	0.14	3	Yes	None
0.1861	0.19	4	Yes	Yes

7.5 Other Specific Types of Impurities

- Chiral Impurities
  Stereoisomers are molecules with identical constitution but different special configurations of atoms. The stereoisomeric pairs of most interest in pharmaceutical drugs are enantiomers. They contain one or more asymmetric (chiral) centers and are nonsuperimposable mirror images. The enantiomers have identical chemical and physical (except for optical rotation) properties. They may have different pharmacokinetic properties (absorption; distribution, biotransformation, and excretion) and pharmacologic or toxicologic effects [5]. In some cases, one enantiomer may not be active and thus has no therapeutic benefit. In either case, the presence and limit for the undesired or inactive enantiomer have to be monitored and controlled.
- Genotoxic Impurities
  Due to high risk associated with such impurities, attempts should be made to prevent their formation including selection of different synthetic routes or mechanisms. If this is not feasible, the safety and appropriate limits for such impurities must be established. ICH provides general guidelines for genotoxic and carcinogenic impurities [6]. However, it is recommended to set more specific limits based on several factors including daily exposure limit and additional studies to support such specifications [7–9].
- Polymorphic Forms
  The polymorphic forms of drug substances defined as follows can have different chemical and physical properties [6, 10]. These differences have potential effect on the quality, safety, efficacy, and the stability of drug products [11]. Therefore, it is recommended to identify and quantitate the polymorphic forms of drug substance in dosage forms [12].
  7) Crystalline forms have different arrangements and/or conformations of the molecules in the crystal lattice.
  8) Amorphous forms consist of disordered arrangements of molecules that do not possess a distinguishable crystal lattice.
  9) Solvates are crystal forms containing either stoichiometric or nonstoichiometric amounts of a solvent. If the incorporated solvent is water, the solvate is commonly known as a hydrate.
- Heavy Metals/Elemental Impurities
  These impurities were classified as a subset of inorganic impurities and discussed briefly in Section 7.3. Due to their toxicity and recent developments for more accurate quantitation of such impurities, this topic is described as follows in more detail.
  The heavy metals of great concern are lead, arsenic, mercury, and cadmium [13]. However, 14 metals are reported as catalysts in the synthesis of pharmaceuticals [14].

Heavy metals/elemental impurities are of great concern for several reasons including metal toxicity and catalysis resulting in the formation of metal complexes and degradation products [15]. Elemental impurities may be present in drug substances, excipients, and drug products. Therefore, their presence and limits have to be reported and controlled.

The current test in the United States Pharmacopeia (USP) for control of heavy metals was introduced in 1905 and is based on wet chemistry procedures involving sulfide precipitation of metals and visual comparison to lead standards [16]. These procedures generate toxic hydrogen sulfide (H_2S). The test is also not sensitive or specific. In addition, it is not reproducible and often underestimates the concentration of several metals [16, 17]. Both European Pharmacopeia (EP) and Japanese Pharmacopeia (JP) have similar procedures for the determination of heavy metals with similar issues regarding sensitivity, selectivity, and reproducibility [18, 19]. Due to these deficiencies, attempts are in progress to replace the existing procedures with more sensitive, selective, and quantitative tests. USP has recently introduced two new general chapters to replace the current procedure. These include General Chapter <232>Elemental Impurities – Limits and General Chapter <233>Elemental Impurities – Procedures [20, 21]. The General Chapter <232>applies to drug products and not drug substances and excipients. However, the limits in drug substances and excipients have to be known and reported. In addition, this chapter does not apply to dietary supplements and veterinary products.

The limits are based on routes of exposure (oral, parenteral, and inhalational) and based on daily dose permissible daily exposure (PDE) of the elemental impurities for drug products. The options for the determination of limits include the following:

Drug Product Analysis Option

The dosage form is analyzed and the results, scaled to a maximum daily dose, are compared to acceptable daily dose PDE.

Summation Analysis Option

The amounts of each elemental impurity in each of the components of the drug product are added, and the results of the summation of each impurity are compared to daily dose PDE limits.

The procedures for evaluation of elemental impurities are highlighted in General Chapter <233>. The two analytical procedures described as reference are based on inductively coupled plasma–atomic (optical) emission (ICP-AES or ICP-OES) or inductively coupled plasma–mass spectrometry (ICP-MS). Alternative procedures are allowed provided that these are validated and meet specified system suitability requirements.

7.6 Non-Drug-Related Impurities

- Water
  Water is a common and unique impurity in drug substances and products. It is unique because this type of impurity also affects the stability and performance of drug products.
  Water can be present as loose molecule (water of adsorption) or tightly bound (water of hydration). In either case, the presence of water may lead to drug chemical instability. The most common example is degradation through hydrolysis or change of drug crystallinity resulting in less stable dosage form or different dissolution rate [22]. The moisture content may also affect other physical properties of dosage form such as hardness and porosity [23, 24]. Therefore, the type and amount of water in dosage forms should be evaluated and controlled. In pharmacopeias, the water content is a common test in drug substance monographs. However, this test is not typically included in the monograph for the drug product. This is due to that fact that the water content has already been monitored and controlled in the corresponding drug substance and excipient used in the preparation of the drug product.
- Extractables and Leachables
  Extractables (potential leachables) are chemical entities, both organic and inorganic, that can be extracted from components of a container closure system into solvents under laboratory experimental conditions [25]. Leachables are chemical entities, both organic and inorganic, that migrate from components of a container closure system into a drug product. FDA has provided guidance for the control of both types of impurities in various documents [26–28].

7.7 Other Sources of Impurities

- Contamination
  These impurities can be introduced into the final product due to contamination or cross-contamination in the manufacturing process. To address this issue, the manufacturers need to follow the current good manufacturing practice (cGMP) guidance and other relevant regulatory documents including internal quality system and standard operating procedures.
- Adulteration
  In spite of all the regulations and quality systems, the introduction of economically motivated adulterated and counterfeited materials into the pharmaceutical supply chain is a recurring problem. Development of more modern, specific, and sensitive analytical techniques for their identification and quantification combined with more effective regulatory enforcements worldwide can minimize such incidents.

7.8 Degradation/Stability Studies

One reason for degradation (stressed) studies is to develop and evaluate stability-indicating assays by demonstrating that the analytical procedure is capable of separating and quantitating potential degradation products generated by stressed conditions (hydrolysis, oxidation, photolysis, thermal, humidity) [29]. The specific conditions for these parameters were discussed in Section 5.9. The other important reason for such studies is to determine the stability of the product by establishing potential degradation and thus impurity profile for the drug substances and products.

The stability studies (long-term and accelerated) have milder conditions than stressed conditions. However, they represent conditions under which more likely degradation products are expected. Therefore, the results of stability studies provide very useful information about impurity profile and degradation pathways. These would help the manufacturers to develop appropriate conditions for storage to minimize degradation products. In addition, they provide insights into identifying, monitoring, and limiting the actual and potential impurities during the shelf life of the products.

7.9 Summary

Different types and sources of impurities in drug substances and products were reviewed. The impurity thresholds and limits of impurities based on general ICH guidelines were discussed. Due to the high risk associated with some selective impurities, more specific and lower limits must be established for qualification of such impurities based on additional supporting safety data.

References

1 ICH, Q3A (R2) Impurities in New Drug Substances, 25 October 2006.
2 ICH, Q3B (R2) Impurities in New Drug Products, 2 June 2006.
3 ICH, Q3C (R5) Guideline for Residual Solvents, February 2011.
4 General Chapter <467> Residual solvents, USP 35-NF 21, United States Pharmacopeial Convention, Inc., Rockville, MD.
5 Development of New Stereoisomeric Drugs, Publication Date: 5/1/1992, http://www.fda.gove/drugss/GuidanceComplianceRegulatoryInformation/Guidances/ucm122883.htm.
6 Guidance for Industry, Q6A *Specifications: Test Procedures and Acceptance Criteria for New Drug Substances and New Drug Products: Chemical Substances,* International Conference on Harmonisation (ICH), 2000.

7 Guidance for Industry Genotoxic and Carcinogenic Impurities in Drug Substances and Products: Recommended Approaches, Food and Drug Administration http://www.fda.gov/cder/guidance/index.htm.
8 Bos PMJ, Baars B, Marcel TM and van Raaij MTM. Risk assessment of peak exposure to genotoxic carcinogens: a pragmatic approach. Toxicol Lett 151 [2004]; 43–50.
9 Question & Answers on the CHMP Guideline on the Limits of Genotoxic Impurities, June 2008, http://www.emea.europa.eu.
10 Byrn SR, Pfeiffer RR and Stowell JG. *Solid-State Chemistry of Drugs.* 2nd Edition, West Lafayette, Indiana: SSCI, Inc., 1999.
11 Brittain H. Methods for the characterization of polymorphs and solvates. In Brittain HG (ed.) Polymorphism in Pharmaceutical Solids. New York: Marcel Dekker, Inc., 1999, pp. 227–278.
12 Guidance for Industry ANDAs: Pharmaceutical Solid Polymorphism Chemistry, Manufacturing, and Controls Information. Division of Drug Information, HFD-240 Center for Drug Evaluation and Research, Food and Drug Administration Rockville, MD. http://www.fda.gov/cder/guidance/index.htm.
13 Q3D: Impurities: Guideline for Metal Impurities, 2009.
14 EMEA Guideline on the Specification Limits for Residues of Metal Catalysis (CPMP/SWP/4446/00).
15 Waterman K. C., Adami R. C., Alsante K. M., Hong J., Landis M.S., Lombardo F. and Roberts C. J. Stabilization of pharmaceuticals to oxidative degradation. Pharm Dev Technol 7 [2002]; 1–31.
16 General Chapter <231>Heavy Metals, USP 35-NF 30, United States Pharmacopeial Convention, Inc., Rockville, MD.
17 Lewen N. et al. A rapid ICP-MS screen for heavy metals in pharmaceutical compounds. J Pharma Biomed Anal 35 [2004]; 739–752.
18 Chapter 2.4. 8. Heavy Metals, 2.4.8, European Pharmacopoeia, 2013 Edition.
19 Chapter 1.07 Heavy Metals Limit Test, The Japanese Pharmacopoeia, Sixteenth Edition, 2011.
20 General Chapter <232>Elemental Impurities – Limits, USP 35-NF 30, United States Pharmacopeial Convention, Inc., Rockville, MD.
21 General Chapter <233>Elemental Impurities – Procedures, USP 35-NF 30, United States Pharmacopeial Convention, Inc., Rockville, MD.
22 Satinder Ahuja and Karen Milles Alsante, Handbook of Isolation and Characterization of Impurities in Pharmaceuticals, 2003, Academic Press, 76.
23 Schepky G. and Fischer M. Effect of production-related variations in hardness and moisture content on the sorption characteristics of tablets. Eur J Pharm Biopharm 39 [1993]; 53–60.
24 Nygvist H., Nichlasson M. and Lundgren P. Studies on the physical properties of tablets and tablet excipients lll: water sorption and its effect on hardness and disintegration. Acta Pharm Suec 18 [1981]; 305–314.

25 Jon S. Kauffman, Identification and risk-assessment of extractables and leachables. Pharm Technol 1 [Feb 2006].

26 Guidance for Industry, Container Closure Systems for Packaging, Human Drugs and Biologics, U.S. Department of Health and Human Services, Food and Drug Administration, May 1999.

27 *Code of Federal Regulations, Title 21,* (Office of Federal Register, Washington, DC, April 2005), Part 211.94, p. 128.

28 FDA, Guidance for Industry: Metered Dose Inhaler (MDI) and Dry Powder (DPI) Drug Products, Rockville, MD: FDA, 1998, pp. 1–65.

29 ICH Q1A, Stability Testing of New Drug Substances and Products, 2003.

8

Good Documentation Practices

Afsaneh Motamed-Khorasani

8.1 Scope

In this chapter, we describe the good documentation practices (GDocPs) and explain why it is important in any regulated environment to manufacture any product. We provide the minimum requirements for good documentation, the reason why it can help, and the rules you need to know and abide by when dealing with documentation in a regulated environment.

GDocPs have been discussed in many different locations throughout different documents from different sources. In other words, there is no single document that explains it in its entirety and in detail. Title 21 of Code of Federal Regulations (CFR) discusses GDocP in Part 11 [1–3], as it related to electronic signatures and electronic documents. In addition, International Organization for Standardization (ISO) 9001:2015 [4], which deals with the requirements for quality management systems, discusses some aspects of GDocP. Furthermore, International Conference on Harmonization (ICH) Q7 (Chapter 6) [5] provides a summary of documentation and records. On the other hand, "The rules governing medicinal products in the European Union (Vol 4)" [6], along with its latest updates in June 30, 2011, summarizes the GDocP rules from the European Union perspective. Therefore, in this chapter, we have tried to combine all this information in order to provide you with a single resource that can provide a broad perspective of GDocP based on different regulatory bodies' requirements, such as Food and Drug administration (FDA) and European Medicines Agency (EMA).

In this chapter, we first define the GDocP, explain its purpose, and provide the reason why it is important to be followed in the process of documentation. We then review some general rules in these lines, to elaborate on multiple aspects of GDocP in more detail, including handling lab notebooks. Electronic documents and electronic signatures, per 21 CFR, Part 11 [1–3], are also described along with the measures to be taken to fulfill the security of these electronic documents and electronic signatures.

Pharmaceutical Analysis for Small Molecules, First Edition. Edited by Behnam Davani.

The US Pharmacopeia (USP) has also recently developed a chapter on good documentation guidelines, Chapter <1029>, numbered above 1000. The USP chapters numbered above 1000 are mainly informative and are provided solely to provide further information and guidance.

The GDocP principals are mostly the same in the European Union as enforced by EMA and are summarized in a document called "The rules governing medicinal products in the European Union (Volume 4)." [6]

8.2 Definition, Purpose, and Importance

Good documentation practice is commonly abbreviated as GDP. However, in order to differentiate it from good distribution practice, which is also abbreviated as GDP, it is recommended that good documentation practice be abbreviated as GDocP. The definition, purpose, and importance of GDocP are discussed in Sections 8.2.1, 8.2.2, and 8.2.3, respectively.

8.2.1 Definition

GDocP is a term in the pharmaceutical industry that describes standards and best practices on how to create, maintain, and archive documents to remain compliant. It is considered to be a part of Current good manufacturing practices (cGMPs), and while not a law, regulatory bodies inspect against the GDocP guidelines. In cases where companies are not following the GDocP guidelines, they may get comments, observations, 483s, and penalties, depending on the importance of the case.

GDocP regulations apply to all personnel, including permanent and temporary employees, interns, summer students, and consultants, who are somehow involved in the process of manufacturing of the regulated product [7]. Similarly, it applies to all activities related to the manufacturing of the regulated product, including manufacture, testing, packaging, labeling, support, holding, storing, and transportation [7].

In order to fully expand on the definition of GDocP, we need to discuss the definitions for "document," "record," and "documentation." ISO has defined these terms in a very clear way. Before providing the definitions of these three terms by ISO, we provide a brief introduction to ISO and its benefits in Section 8.2.1.1.

8.2.1.1 ISO Definition and Benefits

It started in 1926 under the name of "International Federation of the National Standardizing Association." However, it dissolved during World War II. Fortunately, it reorganized again in 1946 as "International Organization for Standardization" (ISO).

It is a voluntary organization with 162 members, where its members are recognized authorities on standards, and each member represents one country. For example, British Standards Institution (BSI), American National Standards Institute (ANSI), Canadian Standards Association (CSA) are a few of these members in the United Kingdom, the United States, and Canada, respectively [8].

Now, let us discuss some of the benefits of ISO. First, it is an assurance that the product and services have a certain level of safety, reliability, and good quality. When you see the ISO mark, irrelevant to the country of manufacture, company, or any other factor, you can be sure of certain minimum reliability and quality of the product. Second, following ISO standards and regulations provides a strategic tool to reduce the cost of manufacturing by minimizing the waste and errors and to increase the productivity. Third, following these standards ensures the accessibility of fast and fair global trade. Therefore, it is a very helpful strategic tool that can assure the customers of a safe, reliable, and high-quality product.

ISO 9000:2005 [9] described the fundamentals of quality management systems and defined related terms. It also provides mutual understanding of the terminology used in quality management (suppliers, customers, and regulators). Right now, ISO 9000:2015 [4] is being followed, but the clauses we used for the definitions of document, record, and documentation were taken from the clauses in ISO 9000:2005 [9].

8.2.1.2 Definition of Document

As indicated in ISO 9000:2005 [9], a "Document" is "Information and its supporting medium." Generally, documents say "do" some activities and explain how to do things. Schematic representation of examples for "Information" and "Media" is presented in Figure 8.1.

8.2.1.3 Definition of Record

As indicated in ISO 9000:2005 [9], a "Record" is "a document stating results achieved or providing evidence of activities performed." Records provide evidence of compliance with established requirements and effectiveness of the operation. Records can be used to document traceability and to provide evidence of verification, preventive action, and corrective action. Generally, records need not be under revision control, and they signify "done" when some activities have been performed. Some of the examples of records are depicted in Figure 8.2.

8.2.1.4 Definition of Documentation

As indicated in ISO 9000:2005 [9], "a set of documents, is frequently called documentation." Specifications and records are examples of "documentation." The main objective of documentation is to introduce sufficient instructional details to facilitate a common understanding of the requirements and to perform sufficient recording of various processes and evaluation of any

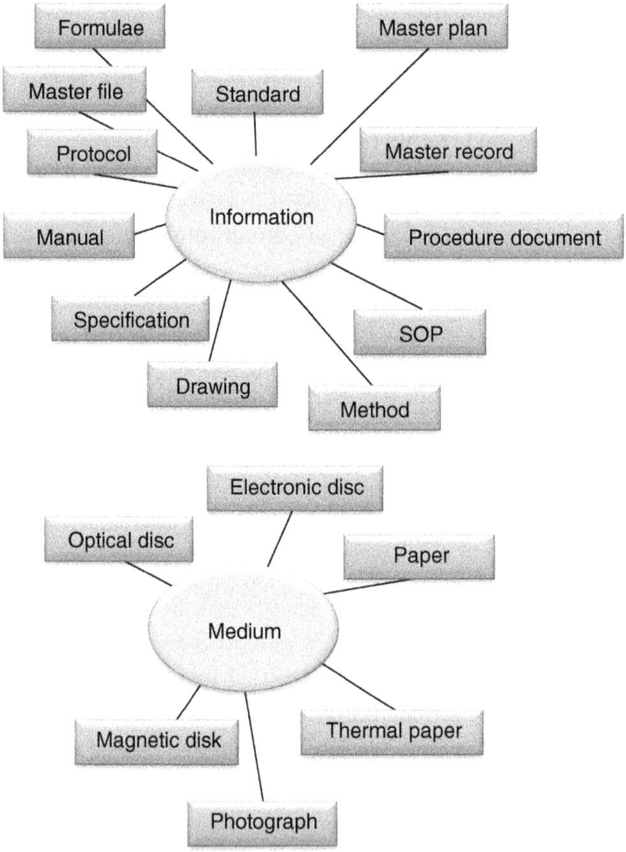

Figure 8.1 Schematic representation of examples for "Information" and "Media."

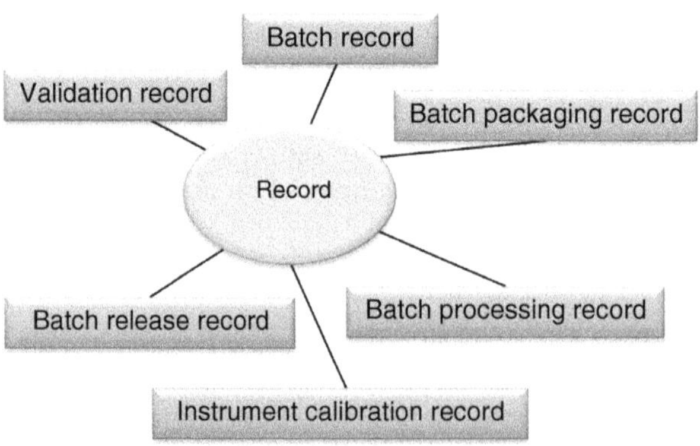

Figure 8.2 Schematic representation of examples for "Record".

observation so that the ongoing application of the requirements could be demonstrated.

There are different types of documents, including instrument printout, notebook, logbook, spreadsheet, data sheet, and manual. Similarly, there are different forms of documentation, including paper, thermal paper, electronic disk, laboratory information management system (LIMS), electronic lab notebook (ELN), magnetic disk, optical computer disk, and photographs.

8.2.2 Purpose of GDocP

As indicated in ISO 9000:2005 [9], "GDocP enables communication of intent and consistency of action." The main objectives of GDocP are to use it as a tool for information transmission and communication in order to communicate the information, to provide evidence of conformity for the provision of evidence that what was planned has actually been done, and to disseminate and preserve the organization's experiences in order to share the knowledge. A typical example would be a technical specification, which can be used as a base for the design and development of a new product.

GDocP ensures that the documents are legible and identifiable, ensures that there are adequate records of all activities, and provides evidence that a product was made according to the regulatory requirements.

These regulations apply to all the steps documents go through from drafting, review, approval, and update. It ensures that identification of the current revision versus the retired version of the documents is a seamless process and prevents the unintended use of obsolete/archived documents. Furthermore, it ensures that there is a process in place to allow for identification of the external documents and controlling their distribution. Finally, it ensures the availability of the current version of the documents at the point of use, in order to minimize the possibility for errors and noncompliance.

In short, GDocP provides detailed instructions and explanation on "what needs to be done (process)," "how to do it (methodology)," "why it needs to be done (context)," "who must do it (responsibilities)," and "when to do it (frequency)."

8.2.3 Importance of GDocP

GDocP is the basic foundation of a quality system to ensure proper documentation and proper control throughout the lifetime of the product. GDocP is essential in a regulated environment to ensure the integrity, traceability, control, and retention of the documents. As FDA puts it, "If it is not written down, it didn't happen." But how it is written is equally important.

GDocP is "expected" as an essential part of quality assurance system in general. It simply increases the chance of product success if GDocP regulations are followed for all aspects of good manufacturing practice (GMP). Following GDocP leads to the accuracy of data and results through maintaining the data integrity and providing correct, complete, current, and consistent information to effectively meet customer/stakeholder's requirements. Furthermore, it creates traceability in all aspects and during the lifetime of a regulated product through facilitation of troubleshooting in case of discrepancies or deviations as well as providing audit trails to be able to address the questions raised during an audit by regulatory bodies.

8.3 General Rules and Principles of GDocP

8.3.1 Requirements of Records

In general, when dealing with records, complete the records as soon as actions are performed. It is against GDocP regulations to fill out the forms or documents before actually performing the tasks, simply because it increases the chance to miss a step or insert wrong information. Once completed, records need to be retained as per applicable retention guidance and for a proper length of time. Furthermore, you need to include appropriate controls to protect the record integrity. Finally, it is very important to create an adequate documentation system that is capable of and optimized for proper archiving and traceability.

As depicted in Figure 8.3, there are eight attributes that records need to have, including the following:

1) Truthful:
   When you sign a record, you are testifying that the information is factual and true.
2) Complete:
   Records should be complete. If you are dealing with a form, all parts need to be filled, and even if it does not apply, place N/A (not applicable) instead of leaving it blank. If there is any part of the page that is unused, cross it with a diagonal line to show that this is unused space, date, and initial.
3) Concise:
   Records need to be concise, capturing only the necessary facts and not subjective guesses or any other extra information.
4) Legible (numbers and characters):
   Records should be legible. What is captured in a record needs to be tidy, organized properly, and easily read if there are handwritten parts in the record. This becomes especially important when you are dealing with some numbers and letters: for example, 0 and 6; U and V; S and 5; 1 and 7; and 3, 8, and B.

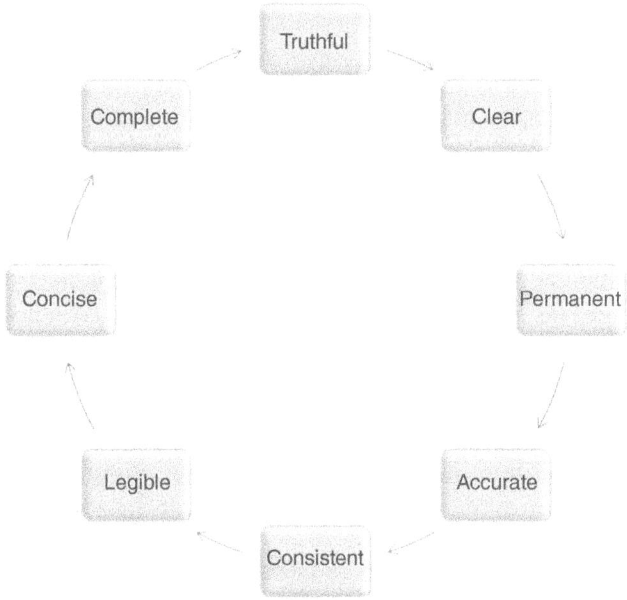

Figure 8.3 Attributes of records based on GDocP regulations.

5) Consistent:
   There should be a good level of consistency throughout the records in terms of formatting for paragraphs, dates, time, styles, numbering, and bullets.
6) Accurate:
   Records should capture the information and facts accurately and detailed enough to enable another person to exactly replicate the experience at a later time. Therefore, all kinds of helpful details need to be included, such as calculations, spellings of chemicals, manufacturer name and address, lot numbers, reference material numbers, serial numbers, and product codes.
7) Permanent:
   The information should be captured in a permanent manner in a record. Therefore, if there is a need to fill out some parts of a printed document, a permanent indelible marker should be used in black. Avoid using any other colors, such as blue, red. Blue color specifically is troublesome at the time of copying the documents and will not show up as a string as when you use a black ink. Avoid using pencils to fill out forms since they are not permanent; and the recorded text can be changed or can fade over time.
8) Clear:
   Records should be clear and easily interpretable by anybody. Avoid using ambiguous terms and interpretations. Be factual, and communicate the observations.

8.3.2 General Tips in GDocP

For handwritten documents, use only black, indelible, ball-point, permanent ink [10–12]. Do not use pencils or nonpermanent markers. Make sure that all the entries are legible, and adequate space is provided to create a clear legible record [10, 11].

All documents should be free from errors [13, 14]. However, in case of any errors (i.e., misspelling, illegible entries, misrepresentation of the data), do not overwrite or scratch and use liquid correction fluid or any masking material [10, 13] Cross a single line through the text that needs correction, write the correct information, the reason for error [10, 13, 15], date, and initial. Do not backdate documents.

While performing a procedure, document each step before moving to the next [10–13]. Do not use ditto marks or continuation lines. If there are any spaces that cannot be filled because they do not apply to what you are doing, do not leave it blank, but put N/A (not applicable) or cross out. Record numbers less than 1 with a 0 before the decimal point. Once complete, sign or initial and date the record.

All documents need to be approved, signed, and dated by an eligible, authorized personnel [10, 15]. However, you cannot approve, verify, or review your own performance, but you need to ask somebody else who is eligible to do so. When a document is electronically produced, the documentation should be checked for accuracy [10]. Please note that a stamp is not acceptable instead of a handwritten signature.

When copying records, the copies made should be legible [10, 11, 16]. You need to be careful not to introduce errors into the document due to the process of copying [10, 14–17].

Regarding document maintenance, documents should be regularly reviewed and updated if needed to be always current [10, 14]. All the documents should be retained for a proper period of time as per regulatory body regulations applicable and be available upon request for review [10, 14–17]. If the documents are being maintained electronically, you need to make sure that the electronic document management system is properly functioning and being validated [14]. In addition, the electronic records should be backed up regularly on safe and reliable media [10, 14]. There should be controls in place to make sure that electronic documents can only be modified and approved by authorized personnel [10, 13]. Access to electronic documents should be controlled by password, identification code, or both, depending on the system being closed or open [10]. The history of all the changes and deletions should be kept (audit trail) [10, 12–14].

Additional expectations can be inferred through extension of the GDocP regulatory guidance. Some of these expectations have been mentioned in this paragraph. The addition of page numbers in the format of "page x of y" allows the reviewer to ensure the completion of the review for all the pages in a document. None of the pages in a document should be removed since it will obscure the data that were present at some time [18].

8.3.2.1 Time Recording

There are two ways to record the time, military time and meridian time, which are described in the following lines [7, 18]. However, the most important things are to always be clear in terms of formatting (sensitivity to cultural differences) and follow your company style when dealing with time recording.

For military time format, 2 digits are used for the hour (00–23) and 2 digits for the minutes (00–59). For example, 09:36 for a time point in the morning and 16:45 for a time point in the afternoon are used.

For meridian time, 1–2 digits are used for the hour (1–12) and 2 digits for the minutes (00–59). In addition, to differentiate morning from an afternoon time, "a.m." and "p.m." are used. For example, 9:36 a.m. for a time point in the morning and 4:45 p.m. for a time point in the afternoon are used.

8.3.2.2 Date Recording

All entries to a GMP document must have a date written on the document, which serves as a tracking system to verify that the task was performed on a certain date [18]. In any case, you need to always be clear in terms of formatting (sensitivity to cultural differences) and follow your company style for entering dates.

All dates should include day, month, and year in a consistent format to avoid confusion. For example, it is best if you represent at least the first three letters of the name of the month. Using the YYYY-MMM-DD format at the start/end of file name is a good format for naming files and folders. Some of the popular formats for capturing the date in your document are as follows:

MM/DD/YYYY	09/23/2009
DDMonYYYY	23Sep2009
DD-Mon-YYYY	23-Sep-2009
Mon DD, YYYY	Sep 23, 2009

8.3.2.3 Backdating

The practice of going back to a previously completed document that has not been properly initiated/dated and adding the dates and initials or placing the date of completion as though filling of the date on which the task was performed in a timely manner is called "backdating." Backdating is not allowed in cGMP environment.

8.3.2.4 Signature and Initial

All entries to a GMP document must have a signature or initials associated with it. It serves as a tracking system to verify if a task was indeed performed and the person who performed it. There is always a meaning for each signature on a regulated document. Based on GDocP, this meaning needs to be clearly mentioned for each signature. A signature can be applied by a performer, verifier, or reviewer, where it means you did the task, you watched it being performed, or you reviewed the data, and it is accurate.

Initials are accepted in most occasions; however, some operations require a signature. Always follow your company style for entering initials and dates. In large companies, a logbook is maintained for the signatures and initials of all the employees.

Now, let us discuss the terms "Performed by, recorded by, and verified by" [7]:

Performed by:

In performing each step of manufacturing, each step of performance should be documented before moving to the next step. Only the personnel who are already trained in the task or are in training under the supervision can initial and date the performance.

Recorded by:

The only time "recorded by" is used instead of "performed by" is when the operator performing a step is unable to initial and date immediately due to working in a confined/restricted area (laminar flow hood). In this case, the data is recorded by another person watching the operation. This person must sign and initial the "recorded by" area of the document.

Verified by:

Verification of each step should be performed prior to the next step. Operators cannot verify their own action (at least one other person must review documentation for accuracy). Verification can be done by the personnel who are already proficient in the task performed and witnessed that a task was performed per written instructions and was documented.

8.3.2.5 Rounding Rules for Numbers

In calculations, the best practices dictate that the extra digits be carried out through the end and then be rounded off. If the removed digit is <5, the preceding digit stays the same (1.874 rounds off to 1.87). If the removed digit is ≥ 5, the preceding digit is increased by 1 (1.875 rounds off to 1.88) [7].

8.3.2.6 Corrections

For an approved printed cGMP document, no handwritten changes are allowed. In case of a need for correction, consult with your supervisor. Any changes required need to be implemented through an established quality system flowchart.

For an approved manually recorded cGMP document, apply a single line through the incorrect information, initial, date beside the crossed-out text, enter the correct data near the original data. The mistake should be still legible through the line (cross-out). Date of the correction is considered to be the date on which the correction was made (not the original date when the wrong data was entered) [7].

8.3.2.7 Missing Data

When the information was not captured at the time of the actual step being performed, mark the blank entry by an asterisk and elsewhere record the same mark and the notation (the data that should have been there). Comments should include the following: the reason why information was missing, proper information that should have been included, and the date of action. The comments should be dated and initialized at the time of recording [7].

In general, one needs to avoid using asterisks as part of the notation of hand change, where sufficient white space is available. Only when there is no sufficient space in the document, a fully notated hand change is permitted, the use of asterisk or any other mark near the correction is a common practice, and elsewhere, one needs to record the same mark and the notation [18].

The use of notation is limited to one per page. The risk with using asterisk or any other mark is that additional changes may be made by other personnel who use the same mark, and the notation could be interpreted to be applicable to all those changes with that specific mark. Therefore, as stated before, if there is enough space in the document to allow entering the correct data, it is best to avoid using asterisk or any other mark for notations [7, 18]. It is best to clearly include the number of changes that the notation applies to, such as "Two entries changed above due to entry of wrong data. AM 14 Feb2016."

There have been no known instances of a regulatory body rejecting a notation with these specifications [18].

8.3.2.8 Voiding Records

In case of discovering errors in making an in-process material (buffer) after the process is complete, one needs to discuss the course of action with the quality assurance manager. If the decision is made to scrap the material, start all over with a new in-process material. The original document should be voided by writing "Void" across the front of the document and be attached to the document replacing it. The approval from the supervisor and quality assurance manager is necessary, and finally, apply your initials and date [7].

8.3.2.9 Recreating and Rewriting of the Records

In general, in a regulated environment, recreating and rewriting of the records should be avoided. In case it is necessary to do this, you can generate records in certain situations only if the original record is illegible or of poor quality, an incorrect form or document was used, the record is physically damaged and not reparable, or the original was in a format that would not keep (thermal paper strips). In any of these cases, you still need the approval from the supervisor and quality assurance manager. Then, identify the recreated document as “rewrite” or “transcript,” and reference the new document to the sources of the information.

Please note that the use of scrap paper and post-it notes and recording raw data on nonofficial records constitute a setup for transcription and are therefore prohibited [7].

8.3.2.10 Deviations

In spite of qualified equipment, trained employees, and validated processes, the daily routine in pharmaceutical manufacturing is prone to situations where the approved workflows fail to deliver the expected results. For this reason, it is important not only to define the workflows for production in "normal cases" but also to specify how such deviations are to be handled. This will ensure the maintenance of the acceptable product quality even in the event of unforeseen circumstances.

Different terms are used for deviation, including discrepancy, atypical situation, and nonconformity. Furthermore, there is no clear, sharply outlined definition in various regulatory documents in the United States or in the European Union (EU). Therefore, it is imperative for a company to internally define what is considered deviation, in order to avoid vagueness. In general, deviations represent a failure to meet specifications (such as parameter settings) in the production process, in-process specifications, or production requirements.

When deviations are being handled, interfaces are created with corrective and preventive actions (CAPA) system and the quality risk management system. Deviations are also taken into account in the Management Review as indicators of how stable processes and workflows are.

Deviations can occur in different areas:

A) Production process:
   a) Manufacturing formula
   b) Process parameters (e.g., machine parameters)
   c) Process specifications (e.g., target values in production process or yield limits)
   d) Testing instructions for in-process controls (e.g., using obsolete versions)
   e) In-process specifications
   f) Anomalies in the process
B) Machines, plants, equipment, facilities, and media (including laboratory):
   a) Machine defects
   b) System failures
   c) Temperature, humidity, number of particles, or pressure differences outside of limits

    d) Deviations in microbiological monitoring
    e) Calibration results outside of limits
    f) Failure to keep calibration or maintenance intervals
C) Regulations:
    a) Standard operating procedure (SOPs)
D) Quality control:
    a) Results out of specification (OOS)
    b) Results out of trend (OOT)
    c) Results close to specification limit
    d) Using expired reference standards
E) Storage:
    a) Anomalies in pest-control sector
    b) Exceeding temperature limits
    c) Anomalies in goods received procedures

In order to address the deviations:

A) Thoroughly record the deviation with all the details.
B) Investigate the potential causes with the details.
C) Assess the risk of deviation for the current batch as well as other batches in the production line that might be affected.
D) Identify and specify suitable actions for affected batches.
E) Specify actions to prevent potential recurrence (corrective actions).
F) Assess the potential risk for the intended corrective actions.
G) Define specific tests for effectiveness/suitability of the proposed corrective actions.
H) Implement the proposed corrective actions.
I) Test the effectiveness and suitability of the proposed corrective actions.
J) Periodically review the system effectiveness.

8.4 General Tips for Laboratory Notebook Documentation

Each company should establish a laboratory notebook policy, a record retention policy, along with documented records for personnel training in lab notebook procedures.

8.4.1 Assignment

Lab notebooks must be assigned a unique identifier number when checked out. They are considered confidential and are the property of the company. When you assign a notebook, list the name of the company, the name of the person, the unique identification number allocated for the notebook, and the date. A schematic representation of hardcover lab notebook is presented in Figure 8.4.

8.4.2 Documentation

In this section, some of the general laboratory notebook documentation tips are included [7].

Use only a permanent ball-point pen in black. Do not use pencil or color markers. Do not use white-outs for correction. It is best practice to start each day's work on a separate page. Once an experiment is completed, draw a diagonal line across the unused portion of the last page, sign, date, and ask your supervisor or the lab manager to sign (as a verifier). If the work is continued on the second page, start the page with the phrase "continued from page xxx" for ease of tracking. Each page must show the date of entry. Similarly, each page must be signed and dated by the person who does the work as well as by a verifier.

When you start writing in the lab notebook, state the title, objective, materials and methods, results and conclusion for the developmental work that could be potentially patentable. Describe all the materials used. Provide the quantity you used for all the materials in this experiment. Record all the operational details and conditions, such as yield, product name, lot numbers, standards used, reference materials used, suppliers, and expirations dates. Try to avoid opinions or negative comments (always be factual). If there is a mistake, you have to draw a line through the mistake and write the correct entry beside it, initial, and write the reason for correction. It is wrong to cover the incorrect data by several lines or white-out.

If data are not kept on the notebook, they must be checked, signed, and dated. In addition, they need to be identified to provide a reference back to a specific page in the lab notebook.

It happens sometimes that you need to attach graphs or charts that are printouts of an instrument. For attachment of graphs or charts to the notebook pages, use permanent adhesives, and date and sign the attachment. The signature should be provided such that it covers both the margin of the attachment and the original page. When unfolded, the attached documents should be within the confines of the opened notebook (not larger). No entry should be made beneath the attached sheets, and no data should be obscured by this attachment. This insert should be checked and dated by a verifier as well. For the sake of clarity of traceability, number all the tables and graphs (including any attachments) in the lab notebook (Table X, Figure Y). If there is a need to explain and elaborate on the results in the table or figure, you can do so underneath the title of the table or figure.

If you need to reference the notebook, include the researcher's initial, notebook number, and notebook page.

Figure 8.4 Schematic representation of a hardcover lab notebook.

8.4.3 Storage

Multiple completed notebooks should not be kept at the researcher's work area. The completed lab notebooks should be kept in a separate room in a safe, fire-resistant metal cabinet that can be locked, where only certain authorized personnel have access to it. They should not be copied without the supervisor's permission. If the personnel need to access the completed lab notebooks at any time, they need to get the permission of the supervisor and check the notebook out and check it back in after reviewing.

8.5 Electronic Documents and Electronic Signatures (21 CFR, Part 11)

In this section of the book, the CFR is defined in order to elaborate on the use of electronic documents and electronic signatures (21 CFR, Part 11) [1–3].

8.5.1 Definition of 21 CFR

The CFR [19] is the codification of the general and permanent rules and regulations that is sometimes called the administrative law. It is published in the Federal Register by the executive departments and agencies of the US federal government by the Office of the Federal Register, an agency of the National Archives and Records Administration (NARA).

CFR is divided into 50 titles that represent broad areas subject to federal regulation. Every regulation in the CFR must have an "enabling statute" or statutory authority that is legally binding. These 50 subject-matter titles contain one or more individual volumes, which are updated once each calendar year, on a staggered basis.

The United States Code (US Code) was started in 1938 and precedes the CFR and contains statutes enacted by the Congress. The US Code is a codification of legislation. The two documents represent different stages in the legislative process. The CFR contains regulations, which spell out in further detail how the executive branch will interpret the law. Therefore, the CFR serves as administrative law.

GDocPs have been discussed in many different locations throughout the CFR in very close relationship with recordkeeping. Some of these CFRs are as follows:

A) General (electronic records and electronic signatures):
   a) 21 CFR Part 11 [1–3]
B) Nonclinical laboratory studies:
   a) 21 CFR Part 58 [20–24]

C) Pharmaceuticals (drugs):
   a) 21 CFR Part 212 [25]
   b) 21 CFR Part 211 [26–28]
   c) 21 CFR Part 312 [29]
D) Medical devices:
   a) 21 CFR Part 812 [30]

There are also other sources that discuss GDocP. Some of these resources are as follows:

E) ISO requirements [8]:
   a) ISO 9000:
      i) A series of standards that deals with the fundamentals of quality management systems, developed and published by ISO.
      ii) It defines, establishes, and maintains an effective quality assurance system for manufacturing and service industries.
   b) ISO 9001:
      i) Deals with the requirements for quality management systems.

F) USP (Chapter <1029>)
G) ICH Q7 [12]:
   a) Chapter 6 provides a summary of documentation and records.
H) EU [6]:
   a) Rules governing medicinal products in the European Union (Vol 4).
   b) There have been some updates in these rules as of June 30, 2011.
   c) These updates could be found in Volume 4 of GMP for Medicinal Products for Human and Veterinary Use (Vol 4).

8.5.2 21 CFR – Subchapter A – General

8.5.2.1 Part 11 – Electronic Records and Electronic Signatures

21 CFR Part 11 consists of three subparts: subpart A (contains the general provisions and describes the code's scope, implementation, and definitions), subpart B (defines the requirements for electronic records), and subpart C (does the same for electronic signatures). In the following sections, all three subparts are elaborated in detail [1].

8.5.2.1.1 Subpart A – General Provisions

Scope In terms of scope, this regulation applies to all electronic records "created, modified, maintained, archived, retrieved or transmitted" under any FDA documentation requirements and includes "electronic records" submitted to the agency under the requirements of the Federal Food, Drug, and Cosmetic Act and the Public Health Service Act, even if such records are not specifically identified in agency regulations. However, the regulation does not apply to electronically transmitted paper documents.

Once electronic signatures and their associated electronic records meet all the regulation's requirements in this part, electronic records and signatures are considered to be equivalent to full handwritten documents, signatures, initials, and other general signings as required by agency regulations. The regulatory body has the right to inspect any hardware or software used to create, maintain, or store these electronic documents.

The regulations in this part set forth the criteria under which the agency considers electronic records, electronic signatures, and handwritten signatures executed to be trustworthy, reliable, and generally equivalent to paper records and handwritten signatures executed on paper.

In addition, computer systems (hardware and software), controls, and attendant documentation maintained under this part shall be readily available for, and subject to, FDA inspection

Implementation For the records required to be maintained but not submitted to the agency, persons may use electronic records in lieu of paper records or electronic signatures in lieu of traditional signatures, in whole or in part, provided that the requirements of this part are met.

For records submitted to the agency, persons may use electronic records in lieu of paper records or electronic signatures in lieu of traditional signatures, in whole or in part, provided that the requirements of this part are met, and the document or parts of a document to be submitted have been identified in public docket No. 92S-0251 as being the type of submission the agency accepts in electronic form. This docket will identify specifically what types of documents are acceptable for submission in electronic form without paper records and the agency receiving unit(s) to which such submissions may be made. Documents sent to the agency receiving unit(s) not specified in the public docket will not be considered as official if they are submitted in electronic form. Paper forms of such documents will be considered as official and must accompany any electronic records.

Definitions In this section, the definition of some of the words that are going to be used in the next few sections are provided.

Handwritten signature: Scripted name or legal mark of an individual handwritten by that individual and executed or adopted with the present intention to authenticate a writing in a permanent form. The act of signing with a writing/marking instrument such as a pen or stylus is preserved. The scripted name/legal mark, while conventionally applied to paper, may also be applied to other devices that capture the name or mark.

Electronic signature: A computer data compilation of any symbol (or series of symbols) executed, adopted, or authorized by an individual to be the legally binding equivalent of the individual's handwritten signature.

Digital signature: An electronic signature based upon cryptographic methods of originator authentication, computed by using a set of rules and parameters such that the identity of the signer and the integrity of the data can be verified.

Biometrics: A method of verifying an individual's identity based on measurement of the individual's physical feature(s) or repeatable action(s), where those features and/or actions are both unique to that individual and measurable.

Electronic record: Any combination of text, graphics, data, audio, pictorial, or other information representation in digital form that is created, modified, maintained, archived, retrieved, or distributed by a computer system.

Open system: An environment in which system access is not controlled by persons who are responsible for the content of electronic records that are on the system.

Closed system: An environment in which system access is controlled by persons who are responsible for the content of electronic records that are on the system.

8.5.2.1.2 Subpart B – Electronic Records

Controls for Closed Systems There should be procedures/controls to be employed to ensure authenticity, integrity, confidentiality of e-records. The signer cannot readily repudiate the signed record as not genuine. These controls/procedures should also ensure the ability to generate accurate/complete copies of records in human-readable form and e-form suitable for inspection, review, and copying by the agency along with protection of records for accurate/ready retrieval in the retention period while limiting system access to authorized individuals [2].

The closed system should possess a system documentation control/procedure for distribution, access, and use of documentation for system operation and maintenance. In addition, there is a need for revision and change of control procedures to document the audit trail for time-sequenced developments and modifications.

The closed system should possess a system validation control/procedure to ensure of accuracy, reliability, consistent intended performance, and ability to discern invalid or altered records.

There is a need to have secure, computer-generated time-stamped audit trails in a closed system in order to be able to independently record the date and time of operator entries along with the actions that create, modify, or delete records, such that record changes do not obscure previously recorded information. Such audit trail documentation should be retained for a certain period of time, depending on the regulations set forth by the applicable regulatory body, and be available for agency review and copying.

Controls/procedures to check the devices and terminals are a necessity to determine the validity of the data input source or operational instruction. Similarly, there should be controls/procedures in place to enable authority checks for using the system, electronically signing a record, accessing the operation or computer system input/output device, altering a record, and performing the operation at hand.

Controls for Open Systems For the open systems, all the controls/procedures discussed for closed system need to be present. Additionally, there need to be controls/procedures in place to enable document encryption along with appropriate digital signature standards to ensure record authenticity, integrity, and confidentiality.

Signature Manifestation Section 11.50 covers signature manifestations, which must include "the printed name of the signer, the date and time when the signature was executed, and the meaning (such as review, approval, responsibility, and authorship) associated with the signature and these items shall be included in any human-readable form of the record."

Specifically, each operator must indicate the intent when signing something, and he or she has to re-enter the user ID and/or password (that shows awareness that he or she is executing a signature) and give the meaning for the electronic record (such as electronic display or printout).

Additionally, signature manifestations must meet all electronic record requirements and are subject to the same controls as for electronic records.

Signature and Record Linking Section 11.70 requires that a given system must link a signature, whether electronic or handwritten, to a particular electronic record in such a manner that signatures are protected from excision, duplication, or transfer that could result in document falsification. If an individual handwrites a signature on an electronically generated document, it must link to the electronic record.

Peripheral but essential data describing the electronic data of interest are called "metadata" and must be integrated into the document they describe.

Metadata might include who owns the data, the author, the size in bytes, and the creation date. The FDA asserts that this link must be technology-based and verifiable and that administrative and procedural controls alone will not protect the document's integrity; however, the FDA does not endorse or require the use of any particular technology to do so.

8.5.2.1.3 *Subpart C – Electronic Signatures*

Each electronic signature shall be unique to one individual and is not to be reused by, or reassigned to, anyone else. Before an organization establishes, assigns, certifies, or otherwise sanctions an individual's electronic signature, the organization shall verify the identity of the individual [3].

Persons using electronic signatures shall, prior to or at the time of such use, certify to the agency that the electronic signatures in their system are intended to be the legally binding equivalent of traditional handwritten signatures. The certification shall be submitted in paper form and signed with a traditional handwritten signature. In addition, upon agency request, the person should be able to provide additional certification or testimony that a specific electronic signature is the legally binding equivalent of the signer's handwritten signature.

Electronic signatures based upon biometrics shall be designed to ensure that they cannot be used by anyone other than their genuine owners.

Electronic Signature Components and Controls Electronic signatures that are based upon biometrics should be designed to ensure that they cannot be used by anyone other than their owners.

Electronic signatures that are not based upon biometrics should employ at least two distinct identification components (identification code and password), should be used only by their genuine owners, and need to be administered and executed to ensure that attempted use of an individual's electronic signature by anyone other than its genuine owner requires collaboration of two or more individuals.

When an individual executes a series of signings during a single, continuous period of controlled system access, the first signing shall be executed using all the electronic signature components, while subsequent signings shall be executed using at least one electronic signature component that is only executable by, and designed to be used only by, the individual.

When an individual executes one or more signings, not performed during a single, continuous period of controlled system access, each signing shall be executed using all of the electronic signature components

Controls for Identification Codes and Passwords In case of using electronic signatures based upon the use of identification codes in combination with passwords should use controls in order to ensure their security and integrity.

Now, let us discuss some of the requirements these controls need to possess. It is important to maintain the uniqueness of the combined identification code/password such that no two individuals have the same combination of identification code/password. In addition, one needs to ensure that identification code/password issuances are checked, recalled, and revised periodically. Furthermore, it is important to follow the loss management procedures in order to electronically deauthorize the lost, stolen, missing, or potentially compromised tokens, cards, and other devices that generate identification code/password information. This should also cover the issuance of temporary or permanent replacements using suitable, rigorous controls.

There should be some transaction safeguards in order to prevent unauthorized use of passwords/identification codes and to detect and report, in an immediate/urgent manner, any attempts at their unauthorized use to the system security unit/organizational management. Initial and periodic testing of the devices (along with the tokens or cards used with them) that generate identification code/password information is also a must, in order to ensure that they function properly and that they have not been altered in an unauthorized manner.

8.6 US Pharmacopeia General Chapter <1029>

8.6.1 Background

In accordance with Section 6 of the Rules and Procedures of the 2010–2015 Council of Experts at USP, an expert panel was proposed to gather important guidelines for good documentation. The panel's inception was on December 01, 2012, with representatives from big and small transnational companies along with representatives from US FDA. They proposed the USP chapter <1029> and proposed the name of the chapter to be "Good Documentation Guidelines."

8.6.2 Purpose

The main purpose of this expert panel was to create a new USP chapter numbered above 1000 to provide information on GDocP to the pharmaceutical industry based on observations/findings of the USP verification program. Some of the potential applications of this chapter are generation, review, approval, and execution of controlled documentation, including but not limited to procedures, protocols, records, and reports. Furthermore, this would create a central document for participants and all industry for GDocP consistency.

8.6.3 Outline of the Chapter

USP Chapter <1029> is a short chapter of a few pages outlining the purpose, scope, principles of good documentation, data collection and recording, and different types of GMP documents, including laboratory records, equipment-related documentation, investigations and deviations, batch records, certificate of analysis (C of A), SOP, protocols and reports, analytical methods, training documentation, and retention of documents.

8.7 Rules Governing Medicinal Products in the European Union (Vol. 4: Documentation)

Directive 2003/94/EC, Volume 4 provides guidance for the interpretation of the principles and guidelines of GMP for medicinal products for human use. In Chapter 4 of this document, documentation is discussed. This document was revised on June 30, 2011. The reason for this revision was to incorporate more of the relevant documentation aspects in relation to the increasing use of electronic documents within the GMP environment. The sections revised included "generation and control of documentation" and "retention of documents" [6].

8.7.1 What is New in the Latest Version?

In this section, a list of new additions to the new revision, as of June 30, 2011, is presented [6]:

- Increased coverage of the use of computer systems.
  - A) First time of mentioning "site master file (SMF)":
    - a) What to include in it?
    - b) The requirement to keep it up-to-date.
    - c) How to submit it as part of site approval process?
  - B) First time of mentioning "standard operating procedure or SOP."
  - C) First time of mentioning that you have to follow your procedure:
    - a) It was not previously mentioned in Chapter 4 that you have to follow your procedures (except for Chapter 1, 5).
  - D) First time mentioning of a need for "imperative mandatory style" for documentation.
  - E) Master list of documents required:
    - a) An inventory of documents within the quality management system should be maintained.
  - F) More clarity on the length of keeping the records:
    - a) The previous version of Chapter 4 mentioned that the records should be kept for 1 year after expiry of the batch.
    - b) Elsewhere in GMP, it was mentioned that "or at least five years after certification of the batch by the qualified person, whichever is longer."
    - c) In the new update, these two sentences were brought together.
  - G) More clarity of records associated with the manufacture of clinical trial materials:
    - a) The records should be kept at least 5 years after the completion or formal discontinuation of the last clinical trial where the batch was used.

H) Increased requirements for areas of policies, procedures, and records:
  a) Technology transfer
  b) Change control
  c) Investigations into deviations/nonconformances
  d) Internal audits
  e) Product quality review
  f) Supplier audits

8.7.2 Outline of EU GDocP Regulations

The outline of the contents within Chapter 4 of the Directive 2003/94/EC, Volume 4, includes [6] the following:

A) Principle
B) Required GMP documentation
C) Generation and control of documentation
D) GDocP
E) Retention of documents
F) Specifications
G) Manufacturing formula and processing instructions
H) Procedures and records

8.7.2.1 Principle

The principle of GDocP in Chapter 4 of the Directive 2003/94/EC, Volume 4, is quite similar in essence to what was previously discussed in this book chapter for US regulations.

In this section of the outline in Chapter 4 of the Directive 2003/94/EC, Volume 4, it is mentioned that there are two major types of documentation to manage/record GMP compliance, including, instructions (directions, requirements) and records/reports. The appropriate GDocP should be applied depending on the document type.

Appropriate controls/procedures should be in place to ensure the accuracy, integrity, availability, and legibility of the documents. The instruction documents should be error-free and recorded/documented on media from which data could be rendered in a human-readable form [6].

8.7.2.2 Required GMP Documentation (by Type)

This section of Chapter 4 of the Directive 2003/94/EC, Volume 4, discusses the required GMP documentation by type as follows [6]:

A) Site master file:
  a) A document describing the GMP-related activities of the manufacturer.

B) Instructions (directions, or requirements) types:
   a) Specifications:
      i) Describe in detail the requirements with which the products/materials used or obtained during manufacture have to conform.
      ii) Serve as a basis for quality evaluation.
   b) Manufacturing formulae processing, packaging, testing instructions:
      i) Provide details on the starting materials, equipment, and computerized systems to be used.
      ii) Specify processing, packaging, sampling, testing instructions.
      iii) In-process controls/process analytical technologies to be employed should be specified with the acceptance criteria.
   c) Procedures (SOPs):
      i) Give directions for performing certain operations.
   d) Protocols:
      i) Give instructions for performing/recording certain discreet operations.
   e) Technical Agreements:
      i) Are agreed between contract givers and acceptors for outsourced activities.
C) Record/report type:
   a) Records:
      i) Provide evidence of various actions taken to demonstrate compliance with instructions. activities, events, investigations.
      ii) In the case of manufactured batches, "batch records", a history of each batch of product, including its distribution. Include the raw data that is used to generate other records.
      iii) For electronic records, the regulated users should define which data are to be used as raw data. At least, all data on which quality decisions are based should be defined as raw data.
   b) Certificate of analysis:
      i) Provide a summary of testing results on samples of products or materials together with the evaluation for compliance to a stated specification.
   c) Reports:
      i) Document the conduct of particular exercises, projects, or investigations, together with results, conclusions, or recommendations.

8.7.2.3 Generation and Control of Documentation

This section of Chapter 4 of the Directive 2003/94/EC, Volume 4, discusses the generation and control of documentation [6].

As per Chapter 4 of this directive, "All types of document should be defined and adhered to. The requirements apply equally to all forms of document media types. Complex systems need to be understood, well documented, validated, and adequate controls should be in place. Many documents (instructions and/or records) may exist in hybrid forms, i.e., some elements as electronic and others as paper based. Relationships and control measures for master documents, official copies, data handling and records need to be stated for both hybrid and homogenous systems. Appropriate controls for electronic documents such as templates, forms, and master documents should be implemented. Appropriate controls should be in place to ensure the integrity of the record throughout the retention period.

Documents should be designed, prepared, reviewed, and distributed with care. They should comply with the relevant parts of Product Specification Files, Manufacturing and Marketing Authorization dossiers, as appropriate. The reproduction of working documents from master documents should not allow any error to be introduced through the reproduction process.

Documents containing instructions should be approved, signed and dated by appropriate and authorized persons. Documents should have unambiguous contents and be uniquely identifiable. The effective date should be defined.

Documents containing instructions should be laid out in an orderly fashion and be easy to check. The style and language of documents should fit with their intended use. Standard Operating Procedures, Work Instructions and Methods should be written in an imperative mandatory style.

Documents within the Quality Management System should be regularly reviewed and kept up-to-date.

Documents should not be hand-written; although, where documents require the entry of data, sufficient space should be provided for such entries.

Handwritten entries should be made in clear, legible, indelible way.

Records should be made or completed at the time each action is taken and in such a way that all significant activities concerning the manufacture of medicinal products are traceable.

Any alteration made to the entry on a document should be signed and dated; the alteration should permit the reading of the original information. Where appropriate, the reason for the alteration should be recorded."

8.7.2.4 Good Documentation Practices

This section of Chapter 4 of the Directive 2003/94/EC, Volume 4, discusses GDocP [6].

As per Chapter 4 of this directive, "Handwritten entries should be made in clear, legible, indelible way. Records should be made or completed at the time each action is taken and in such a way that all significant activities concerning the manufacture of medicinal products are traceable. Any alteration made to the entry on a document should be signed and dated; the alteration should permit the reading of the original information. Where appropriate, the reason for the alteration should be recorded."

8.7.2.5 Retention of Documents

This section of Chapter 4 of the Directive 2003/94/EC, Volume 4, discusses retention of documents [6].

As per Chapter 4 of this directive, "It should be clearly defined which record is related to each manufacturing activity and where this record is located. Secure controls must be in place to ensure the integrity of the record throughout the retention period and validated where appropriate. Specific requirements apply to batch documentation which must be kept for one year after expiry of the batch to which it relates or at least five years after certification of the batch by the Qualified Person, whichever is the longer. For investigational medicinal products, the batch documentation must be kept for at least five years after the completion or formal discontinuation of the last clinical trial in which the batch was used. Other requirements for retention of documentation may be described in legislation in relation to specific types of product (e.g. Advanced Therapy Medicinal Products) and specify that longer retention periods be applied to certain documents.

For other types of documentation, the retention period will depend on the business activity which the documentation supports. Critical documentation, including raw data (for example relating to validation or stability), which supports information in the Marketing Authorization should be retained whilst the authorization remains in force. It may be considered acceptable to retire certain documentation (e.g. raw data supporting validation reports or stability reports) where the data has been superseded by a full set of new data. Justification for this should be documented and should take into account the requirements for retention of batch documentation; for example, in the case of process validation data, the accompanying raw data should be retained for a period at least as long as the records for all batches whose release has been supported on the basis of that validation exercise. The following section gives some examples of required documents. The quality management system should describe all documents required to ensure product quality and patient safety."

8.7.2.6 Specifications

This section of Chapter 4 of the Directive 2003/94/EC, Volume 4, discusses specifications [6].

As per Chapter 4 of this directive, "There should be appropriately authorized and dated specifications for starting and packaging materials, and finished products.

A) <u>Specifications for starting and packaging materials</u>
   Specifications for starting and primary or printed packaging materials should include or provide reference to, if applicable:
   a) A description of the materials, including:
      i) The designated name and the internal code reference;

   ii) The reference, if any, to a pharmacopoeial monograph;
   iii) The approved suppliers and, if reasonable, the original producer of the material;
   iv) A specimen of printed materials;
  b) Directions for sampling and testing;
  c) Qualitative and quantitative requirements with acceptance limits;
  d) Storage conditions and precautions;
  e) The maximum period of storage before re-examination.

B) Specifications for intermediate and bulk products
  a) Specifications for intermediate and bulk products should be available for critical steps or if these are purchased or dispatched. The specifications should be similar to specifications for starting materials or for finished products, as appropriate.

C) Specifications for finished products
  a) Specifications for finished products should include or provide reference to:
   i) The designated name of the product and the code reference where applicable;
   ii) The formula;
   iii) A description of the pharmaceutical form and package details;
   iv) Directions for sampling and testing
   v) The qualitative and quantitative requirements, with the acceptance limits;
   vi) The storage conditions and any special handling precautions, where applicable;
   vii) The shelf-life."

8.7.2.7 Manufacturing Formula and Processing Instructions

This section of Chapter 4 of the Directive 2003/94/EC, Volume 4, discusses manufacturing formula and processing instructions [6].

As per Chapter 4 of this directive, "Approved, written Manufacturing Formula and Processing Instructions should exist for each product and batch size to be manufactured.

A) The Manufacturing Formula should include:
  a) The name of the product, with a product reference code relating to its specification;
  b) A description of the pharmaceutical form, strength of the product and batch size;
  c) A list of all starting materials to be used, with the amount of each, described;
  d) Mention should be made of any substance that may disappear in the course of processing;

e) A statement of the expected final yield with the acceptable limits, and of relevant intermediate yields, where applicable

B) The Processing Instructions should include:
   a) A statement of the processing location and the principal equipment to be used;
   b) The methods, or reference to the methods, to be used for preparing the critical equipment (e.g. cleaning, assembling, calibrating, sterilizing);
   c) Checks that the equipment and work station are clear of previous products, documents or materials not required for the planned process, and that equipment is clean and suitable for use;
   d) Detailed stepwise processing instructions [e.g., checks on materials, pre-treatments, sequence for adding materials, critical process parameters (time, temp etc.)];
   e) The instructions for any in-process controls with their limits;
   f) Where necessary, the requirements for bulk storage of the products; including the container, labeling and special storage conditions where applicable;
   g) Any special precautions to be observed.

C) Packaging Instructions:
   a) Approved Packaging Instructions for each product, pack size and type should exist. These should include, or have a reference to, the following:
      i) Name of the product; including the batch number of bulk and finished product
      ii) Description of its pharmaceutical form, and strength where applicable;
      iii) The pack size expressed in terms of the number, weight or volume of the product in the final container;
      iv) A complete list of all the packaging materials required, including quantities, sizes and types, with the code or reference number relating to the specifications of each packaging material;
      v) Where appropriate, an example or reproduction of the relevant printed packaging materials, and specimens indicating where to apply batch number references, and shelf life of the product;
      vi) Checks that the equipment and work station are clear of previous products, documents or materials not required for the planned packaging operations (line clearance), and that equipment is clean and suitable for use.
      vii) Special precautions to be observed, including a careful examination of the area and equipment in order to ascertain the line clearance before operations begin;
      viii) A description of the packaging operation, including any significant subsidiary operations, and equipment to be used;

ix) Details of in-process controls with instructions for sampling and acceptance limits.

D) Batch Processing Record:

b) A Batch Processing Record should be kept for each batch processed. It should be based on the relevant parts of the currently approved Manufacturing Formula and Processing Instructions, and should contain the following information:

i) The name and batch number of the product;
ii) Dates and times of commencement, of significant intermediate stages and of completion of production;
iii) Identification (initials) of the operator(s) who performed each significant step of the process and, where appropriate, the name of any person who checked these operations;
iv) The batch number and/or analytical control number as well as the quantities of each starting material actually weighed (including the batch number and amount of any recovered or reprocessed material added);
v) Any relevant processing operation or event and major equipment used;
vi) A record of the in-process controls and the initials of the person(s) carrying them out, and the results obtained;
vii) The product yield obtained at different and pertinent stages of manufacture;
viii) Notes on special problems including details, with signed authorization for any deviation from the Manufacturing Formula and Processing Instructions;
ix) Approval by the person responsible for the processing operations.

E) Batch Packaging Record:

c) A Batch Packaging Record should be kept for each batch or part batch processed. It should be based on the relevant parts of the Packaging Instructions. The batch packaging record should contain the following information:

i) The name and batch number of the product,
ii) The date(s) and times of the packaging operations;
iii) Identification (initials) of the operator(s) who performed each significant step of the process and, where appropriate, the name of any person who checked these operations;
iv) Records of checks for identity and conformity with the packaging instructions, including the results of in-process controls;
v) Details of the packaging operations carried out, including references to equipment and the packaging lines used;
vi) Whenever possible, samples of printed packaging materials used, including specimens of the batch coding, expiry dating and any additional overprinting;

vii) Notes on any special problems or unusual events including details, with signed authorization for any deviation from the Packaging Instructions;
viii) The quantities and reference number or identification of all printed packaging materials and bulk product issued, used, destroyed or returned to stock and the quantities of obtained product, in order to provide for an adequate reconciliation. Where there are there are robust electronic controls in place during packaging there may be justification for not including this information
ix) Approval by the person responsible for the packaging operations."

8.7.2.8 Procedures and Records

This section of Chapter 4 of the Directive 2003/94/EC, Volume 4, discusses procedures and records [6].

A) "Receipt:
  a) There should be written procedures and records for the receipt of each delivery of each starting material (including bulk, intermediate or finished goods), primary, secondary and printed packaging materials.
  b) The records of the receipts should include:
    i) The name of the material on the delivery note and the containers;
    ii) The "in-house" name and/or code of material (if different from a);
    iii) Date of receipt;
    iv) Supplier's name and, manufacturer's name;
    v) Manufacturer's batch or reference number;
    vi) Total quantity and number of containers received;
    vii) The batch number assigned after receipt;
    viii) Any relevant comment.
  c) There should be written procedures for the internal labeling, quarantine and storage of starting materials, packaging materials and other materials, as appropriate.
B) Sampling:
  a) There should be written procedures for sampling, which include the methods and equipment to be used, the amounts to be taken and any precautions to be observed to avoid contamination of the material or any deterioration in its quality.
C) Testing:
  a) There should be written procedures for testing materials and products at different stages of manufacture, describing the methods and equipment to be used. The tests performed should be recorded.
D) Other:
  a) Written release and rejection procedures should be available for materials and products, and in particular for the certification for sale of the

finished product by the Qualified Person(s). All records should be available to the Qualified Person. A system should be in place to indicate special observations and any changes to critical data.

b) Records should be maintained for the distribution of each batch of a product in order to facilitate recall of any batch, if necessary.

c) There should be written policies, procedures, protocols, reports and the associated records of actions taken or conclusions reached, where appropriate, for the following examples:
   i) Validation and qualification of processes, equipment and systems;
   ii) Equipment assembly and calibration;
   iii) Technology transfer;
   iv) Maintenance, cleaning and sanitation;
   v) Personnel matters including signature lists, training in GMP and technical matters,
   vi) Clothing and hygiene and verification of the effectiveness of training.
   vii) Environmental monitoring;
   viii) Pest control;
   ix) Complaints;
   x) Recalls;
   xi) Returns;
   xii) Change control;
   xiii) Investigations into deviations and non-conformances;
   xiv) Internal quality/GMP compliance audits;
   xv) Summaries of records where appropriate (e.g., product quality review);
   xvi) Supplier audits.

d) Clear operating procedures should be available for major items of manufacturing and test equipment.

e) Logbooks should be kept for major or critical analytical testing, production equipment, and areas where product has been processed. They should be used to record in chronological order, as appropriate, any use of the area, equipment/method, calibrations, maintenance, cleaning or repair operations, including the dates and identity of people who carried these operations out.

f) An inventory of documents within the Quality Management System should be maintained."

8.8 GDocP Enforcement

The regulatory authorities in charge are empowered to inspect establishments in order to enforce the law and its interpretations (guidance document contents).

8.8.1 Regulatory Bodies in Charge

GDocPs are enforced by the regulatory bodies around the world. A list of some of the most popular regulatory bodies and the relative countries is provided in Table 8.1.

8.8.2 FDA GDocP Compliance Observations

Some of the FDA observations on poor documentation practices are listed in this section and include the following: correction of errors in a document without proper signature, date, and reasoning; write-overs, multiple line-through, use of white-out, or other masking devices; lack of documentation for sample sequence tables, lack of authorization of quality-assurance-related SOPs by the quality assurance manager (including production, calibration, storage, and maintenance); lack of recording/documenting the delegate for batch release in case of quality assurance manager's absence; and lack of a detailed procedure, flowchart, and checklist for OOS events.

Table 8.1 List of some of the regulatory bodies around the world and their relative countries.

Country	Regulatory body
United States	Food and Drug Administration (FDA)
European Union	European Commission (EC)
Japan	Ministry of Health, Labor and Welfare (MHLW)
India	Central Drug Standards Control Organization (CDSCO)
China	China Food and Drug Administration (CFDA)
Hong Kong	Pharmaceutical Service Drug Office Department of Health (PSDH) Medical Device Control Office (MDCO)
Korea	Ministry of Food and Drug Safety (MFDS)
Australia	Therapeutic Goods Administration (TGA)
Malaysia	Ministry of Health (MOH)
Philippines	Department of Health (DOH) Philippines Food and Drug Administration (PFDA)
Singapore	Health Sciences Authority (HAS) Health Products Regulation Group (HPRG)
Taiwan	Taiwan Food and Drug Administration (TFDA)
Thailand	Food and Drug Administration Thailand (FDA Thailand)
Vietnam	Ministry of Health (MOH), including: • The Drug Administration of Vietnam (DAV) • Department of Medical Equipment and Health Works (DMEHW)

8.8.3 FDA GDocP Fraud Observations

Some of the FDA observations of fraud that were spotted in the last few years included the following: creating, altering, or deleting data to fit the acceptance criteria; purposefully backdating the documents; backentering data with no traceability (i.e., initialing and dating data entry); signing another person's name; and hiding or throwing away undesired data.

8.8.4 Excerpts of 483 GDocP Observations

In this section, some of the examples where such enforcement has occurred due to GDocP noncompliance are given.

Documentation not contemporaneous

A) US FDA Warning Letter 3201120 (UCM271708) to Yag-Mag Labs Private Limited (Hyderabad, India), September 12, 2011 [31].
B) US FDA Warning Letter UCM172108 to Caraco Pharmaceutical Laboratories, Ltd. (Caraco), May 12, 2009 [32].
C) US FDA Warning Letter UCM076496 to Kunshan Chemical & Pharmaceutical Co., Ltd. (Kunshan City, Jiangsu, China), September 06, 2007 [33].
D) US FDA Warning Letter UCM075472 to Litron Laboratories, Ltd. (Rochester, NY), July 01, 2005 [34].

Use of ditto marks

A) US FDA Warning Letter FLA9929 to All Medicare Home Aids, Inc., January 28, 1999 [35].

Use of signature stamp

A) US FDA Warning Letter UCM075960 to Scott A. Spiro, MD, June 28, 2006 [36].
B) US FDA Warning Letter UCM066113 to Medtronic, Inc., December 02, 1997 [37].

Obscured original data

A) US FDA Warning Letter UCM069041 to SOL Pharmaceuticals Limited, November 21, 2000 [38].
B) US FDA Warning Letter UCM076246 to Gynétics Medical Products NV, January 16, 2007 [39].

Use of pencil

A) US FDA Warning Letter 3200102 to SOL Pharmaceuticals Limited, November 21, 2000 [38].
B) US FDA Warning Letter UCM221006 to Haw Par Healthcare Limited (Singapore), July 20, 2010 [40].

Inaccurate records

A) US FDA Warning Letter 3200102 to SOL Pharmaceuticals Limited, November 21, 2000 [38].

Hand changes not dated

B) Form FDA 483 issued to L. Perrigo Co., November 7, 2008 [41].

8.9 Summary

In summary, in this chapter, we described GDocP and explained why it is important in any regulated environment to manufacture any product. We provided the minimum requirements for good documentation, the reason why it can help, and the rules you need to know and abide by when dealing with documentation in a regulated environment. General rules and regulations of GDocP were covered based on US FDA, EMA, and USP regulations. Furthermore, we covered the general tips specifically for laboratory notebook documentation. Furthermore, we discussed 21 CFR Part 11 dealing with electronic records and electronic signatures. Finally, we provided some excerpts of FDA warning letters for different points discussed through the chapter to provide a real-life example of the problems captured by FDA. We hope that this chapter has been able to provide you with the basics of GDocP regulations with enough clarity and examples that can enable you to understand and follow these guidelines in your work environment in order to save time and eliminate some potential future problems that can be easily prevented.

Abbreviations

ANSI	American National Standards Institute
BSI	British Standards Institution
CAPA	corrective and preventive actions
CFR	Code of Federal Regulations

CSA	Canadian Standards Association
ELN	electronic lab notebook
EU	European Union
FDA	Food and Drug Administration
GDocP	good documentation practice
GMP	good manufacturing practice
ICH	International Conference on Harmonization
IRB	Institutional Review Board
ISO	International Organization for Standardization
LIMS	Lab Information Management System
NARA	National Archives and Records Administration
OOS	results out of specification
OOT	results out of trend
QC	quality control
QMS	quality management system
SOP	standard operating procedures
SMF	site master file
US	The United States
USP	US Pharmacopeia

References

1 FDA. Code of Federal Regulations. Title 21, Chapter I, SubChapter A, Part 11, SubPart A, Section 1–3.
2 FDA. Code of Federal Regulations. Title 21, Chapter I, SubChapter A, Part 11, SubPart B, Section 10–70.
3 FDA. Code of Federal Regulations. Title 21, Chapter I, SubChapter A, Part 11, SubPart C, Section 100–300.
4 ISO 9000:2015. http://www.iso.org/iso/catalogue_detail?csnumber=45481 (accessed 15 November 2016).
5 ICH Q7. Chapter 6.
6 EU Regulations. Volume 4.
7 Nick Kapp. Good documentation practices. http://view.officeapps.live.com/op/view.aspx?src=http%3A%2F%2Fwww.smccd.net%2Faccounts%2Fkapp%2Fbt415%2Fsop%2F1000.01%2520Good%2520documentation%2520Practices.doc (accessed 20 December 2013).
8 ISO. http://www.iso.org/iso/home.html (accessed 15 November 2016).
9 ISO 9000:2005. http://www.iso.org/iso/catalogue_detail?csnumber=42180 (accessed 15 November 2016).
10 EudraLex Volume 4, Good Manufacturing Practice, Medicinal Products for Human and Veterinary Use (Chapter 4: Documentation, Revision 1).

EudraLex. (http://ec.europa.eu/health/files/eudralex/vol4/chapter4_012011_en.pdf).
11 WHO. TRS 961 "Good Manufacturing Practices for Pharmaceutical Products: Main Principles" 2011 (Annex 3, Section 15) (http://whqlibdoc.who.int/trs/WHO_TRS_961_eng.pdf#page=106).
12 ICH. "Q7: Good Manufacturing Practice Guide for Active Pharmaceutical Ingredients" (Section 6) (http://www.ich.org/cache/compo/3632721.html#Q7A).
13 US FDA. Guidance for industry computerized systems used in clinical trials 1999 http://www.fda.gov/ICECI/EnforcementActions/BioresearchMonitoring/ucm135196.htm (accessed 04 February 2010).
14 European Commission Directive 2003/94/EC. (Article 9) (http://ec.europa.eu/enterprise/sectors/pharmaceuticals/files/eudralex/vol1/dir_2003_94/dir_2003_94_en.pdf).
15 21CFR211 Subpart J http://www.accessdata.fda.gov/scripts/cdrh/cfdocs/cfcfr/CFRSearch.cfm?CFRPart=211&showFR=1&subpartNode=21:4.0.1.1.11.10.
16 Japan MHLW. Ministerial Ordinance No. 169, 2004, "Ministerial Ordinance on Standards for Manufacturing Control and Quality Control for Medical Devices and In Vitro Diagnostic Reagents" (Chapter 2, Section 2, Article 8, Paragraphs 2 and 4) (http://www.pmda.go.jp/english/service/pdf/ministerial/050909betsu3.pdf) accessed 07 July 2011.
17 "The rules governing medicinal products in the European Union", EudraLex (pdf), Volume 4, EU Guidelines to Good Manufacturing Practice, Medicinal Products for Human and Veterinary Use., p. 3, Revised 01 June 2006 Check date values in: |date= (help); |chapter= ignored (help).
18 Hurd, Don (2010), Good Documentation Practices (PDF), et al., Premier Validation, pp. 11,17,30,39, ISBN 9781908084002, archived (PDF) from the original on 13 Nov 2012.
19 CFR – Code of Federal Regulations Title 21. https://www.accessdata.fda.gov/scripts/cdrh/cfdocs/cfcfr/CFRSearch.cfm (accessed 15 November 2016).
20 FDA. Code of Federal Regulations. Title 21, Chapter I, SubChapter A, Part 58, SubPart B, Section 81.
21 FDA. Code of Federal Regulations. Title 21, Chapter I, SubChapter A, Part 58, SubPart B, Section 33.
22 FDA. Code of Federal Regulations. Title 21, Chapter I, SubChapter A, Part 58, SubPart D, Section 63.
23 FDA. Code of Federal Regulations. Title 21, Chapter I, SubChapter A, Part 58, SubPart G, Section 130.
24 FDA. Code of Federal Regulations. Title 21, Chapter I, SubChapter A, Part 58, SubPart J, Section 185–195.
25 FDA. Code of Federal Regulations. Title 21, Chapter I, SubChapter A, Part 212, SubPart F, Section 50.

26 FDA. Code of Federal Regulations. Title 21, Chapter I, SubChapter A, Part 211, SubPart B, Section 22.
27 FDA. Code of Federal Regulations. Title 21, Chapter I, SubChapter A, Part 211, SubPart F, Section 100.
28 FDA. Code of Federal Regulations. Title 21, Chapter I, SubChapter A, Part 211, SubPart J, Section 180–198.
29 FDA. Code of Federal Regulations. Title 21, Chapter I, SubChapter A, Part 312, SubPart D, Section 62.
30 FDA. Code of Federal Regulations. Title 21, Chapter I, SubChapter A, Part 812, SubPart G, Section 140.
31 US FDA. "Warning Letter" UCM271708. Observation #1 http://www.fda.gov/ICECI/EnforcementActions/WarningLetters/2011/ucm271708.htm (accessed 27June 2013).
32 Caraco Pharmaceutical Laboratories, Ltd. Response to the FDA's May 12, 2009, 483. Observation #14a http://www.fda.gov/downloads/AboutFDA/CentersOffices/ORA/ORAElectronicReadingRoom/UCM172108.pdf (accessed 01 June 2011).
33 US FDA. "Warning Letter" UCM076496. Observation #1 http://www.fda.gov/ICECI/EnforcementActions/WarningLetters/2007/ucm076496.htm (accessed 16 August 2012).
34 US FDA. "Warning Letter" UCM075472. Observation #4 http://www.fda.gov/ICECI/EnforcementActions/WarningLetters/2005/ucm075472.htm (accessed 16 August 2012).
35 US FDA. "Warning Letter" FLA9929 http://www.fda.gov/downloads/ICECI/EnforcementActions/WarningLetters/1999/UCM067076.pdf.
36 US FDA. "Warning Letter" UCM075960 http://www.fda.gov/ICECI/EnforcementActions/WarningLetters/2006/ucm075960.htm (accessed 04 February 2010).
37 US FDA. "Warning Letter" UCM066113 http://www.fda.gov/downloads/ICECI/EnforcementActions/WarningLetters/1997/UCM066113.pdf.
38 US FDA. "Warning Letter" UCM069041 http://www.fda.gov/downloads/ICECI/EnforcementActions/WarningLetters/2000/UCM069041.pdf.
39 US FDA. "Warning Letter" UCM076246, Observation #13b. http://www.fda.gov/ICECI/EnforcementActions/WarningLetters/2007/ucm076246.htm (accessed 01 June 2011).
40 US FDA. "Warning Letter" UCM221006. Observation #7. b. http://www.fda.gov/ICECI/EnforcementActions/WarningLetters/2010/ucm221006.htm (accessed 16 August 2012).
41 US FDA. Form FDA 483 issued to L. Perrigo Co., dated 11/07/2008. Observation #9A http://www.fda.gov/downloads/AboutFDA/CentersOffices/ORA/ORAElectronicReadingRoom/UCM214731.pdf.

9

The Management of Analytical Laboratories

Roger Dabbah

9.1 Introduction

In general, scientists that work in an analytical laboratory, be it in government, industry, or academia, are not trained in management. There is quite a difference between managing people and managing technical procedures. In addition, few scientists are inclined to want to manage people and all their foibles.

A book on the analysis of small molecules, as well as on large molecules, will need, even if it is only from a defensive perspective, a discussion of the management of these laboratories. The scope of this chapter includes all analytical laboratories, including chemical, physical, and microbiological testing. The mission of an analytical laboratory is to provide the technical assessment of current products, products under development, and products for the future. As such, the laboratory is considered a strategic as well as a tactical component of the organization. This had not been the case in the past, where laboratories were considered as cost centers (i.e., utilizing funds) rather than revenue-generating centers. In addition, in most organizations, the laboratories are not represented well, if at all, in decision-making meetings. The laboratories get attention from senior management only when there are problems to be solved that could affect the bottom line. In my experience, when a problem is identified, the reaction of senior management is to "blame" the inadequacy of the analysis, as if killing of the bearing of bad news will resolve problems.

A number of components in a laboratory will affect the usefulness of analytical results, and these include personnel, facilities, standard operating procedures (SOP), equipment, training of the workforce, internal and external environment, and communications [4]. The integration of these components will be made via a rigorous and progressive management following some general principles of management adapted to the management of laboratories. The overall performance of the laboratories is critical because of the increased scrutiny by the regulatory agencies, such as Food and Drugs Administration (FDA),

Pharmaceutical Analysis for Small Molecules, First Edition. Edited by Behnam Davani.

which demands adherence to cGMPs as well as GLPs in developing the analytical data. The evolution of cGMPs and GLPs requires attention from the laboratories to keep up with these requirements. Nonregulatory agencies such as the United States Pharmacopeia (USP) evolve as technology advances (see Chapter 2), and the analytical procedures indicated are enforceable by FDA in a uniform way, which is applicable to the leveling of the playing field of analysis [6, 9].

9.2 Principles of Management Applicable to the Laboratory Function

9.2.1 System Thinking

This principle is often misunderstood by laboratory personnel that focus on the technical aspect of analytical tasks [1, 2]. The laboratory function in an organization must be integrated with the other functions to accomplish the organization objectives. The contributions of the laboratory, according to the principle of system thinking, need to be optimized over the accomplishments of the organization's objectives. An example could be that the analytical laboratory need not use the most advanced techniques or equipment to gather data, as long as the current methods have been validated and are acceptable to FDA. The most advanced techniques, as elegant as they are, are costly, and from the standpoint of the organization, minimizing the costs takes precedence over advanced techniques.

9.2.2 Organizational Structure

The use of a centralized analytical laboratory in an industrial organization has a number of advantages [2]. It can provide uniform and consistent services to all divisional groups using procedures that have been validated. It also facilitates technology transfer among divisions, thus reducing the cost of each unit of analytical data, based on the principle of the experience curve. The experience curve indicates that when the number of a given assay increases, the cost per assay decreases in a predictable manner. The concentration of analytical "brain power" also promotes better solutions to critical problems encountered in the analysis. The disadvantages of such an organizational structure include a long queue for divisions to obtain priority in the development of analytical data that could be critical.

The use of decentralized analytical function is advantageous because it will give priority to divisional needs. However, if each business unit has its own analytical laboratory, it will have to duplicate the personnel, equipment, and facilities that will increase the cost of obtaining data. Furthermore, effective technology transfer among business units will be reduced with a net loss for the overall organization.

9.2.3 Accountability and Responsibility

Accountability principle requires that each employee in a laboratory when given an analytical task must own that task and is accountable for it success or failure [3]. This requires the individual to understand the procedures to be used, the calibration of equipment, the validation of methodologies, and the reporting of the results in a timely manner. The quality of the data will be developed under GLPs and/or cGMPs such that it would be presentable to regulatory agencies.

Responsibility principle indicates that the individual, once he/she had accepted the assignment, is required to be responsible for it in terms of timeliness, quality, and completeness.

9.2.4 Management of Personnel

In an ideal world, you hire the right people, give them all needed resources, remove all obstacles that could impede their ability to do their assigned tasks, provide a climate that is conducive to innovation and creativity, and reduce their stress. Often, this ideal situation does not exist, and the management of people in a laboratory is influenced by a number of real factors. These factors include the physical design of the laboratory, the diversity of the workforce in terms of education and work ethics.

The manager of analytical laboratories must, of course, have high technical skills commensurate with the type of analysis being performed. However, the manager should also possess soft skills that can provide a seamless management of technical skills blended with soft skills. Among these soft skills are communication, critical thinking, decision-making, time management, problem-solving, and team building [11]. One should not expect that every laboratory manager to have all these soft skills, but enough of them to provide leadership to the laboratory.

9.2.5 Allocation and Utilization of Resources

The resources needed to operate an analytical laboratory, personnel, equipment, facility, and operating funds must be made available to the laboratory for it to accomplish its mission. Resources, in any organization, are generally limited, and allocation of resources is made based on senior management prioritization. The role and responsibility of the management of the laboratory are to provide senior management with a realistic budget. The realistic budget can be divided in two budgets [2]: one for operating funds and the other for capital expenditures. Operating funds consist of routine tasks, salaries, and benefits and of more strategic projects such the development of new products servicing the R&D function. The capital expenditure budget for the laboratory will be incorporated with the capital expenditure budget of the organization and will follow the requirements imposed by the organization for such expenditures (return on investment, payback, amortization schedules with tax implications). Although a laboratory would like to have the latest state-of-the-art equipment, an organization must find alternative ways to accomplish the needed analysis without having to resort to expensive, state-of-the-art equipment.

The utilization of resources must be monitored in order to keep the utilization of resources under control and within the budget. Sometimes, because of unexpected additional regulatory requirements, the analytical laboratory will need additional resources, and these should be requested on a supplemental budget.

9.2.6 Internal Interactions

The analytical laboratory is not an island and thus must communicate with the various functions of the organization to accomplish the organizational objectives in an efficient and effective way. Scientists are known to use a language that is understood only by other scientists. The use of scientific jargon, especially acronyms, becomes an obstacle to communication between the laboratory and the remaining of the organization.

A manager of an analytical laboratory has to develop political skills that sometimes are necessary, especially if the philosophy of the organization is directed toward short-term gains at the expense of long-term plans. In essence, one should know where the "hot" buttons are, where resources are available, and who are the champions that could tilt the balance of resource allocations toward the analytical laboratory.

9.2.7 External Interactions

In the pharmaceutical/biotech industry, the relationship between the organization and the regulatory agencies, such as FDA, is critical. The analytical laboratory function participates in the development of new products or modified products, and the data it generates is submitted to the FDA as part of the regulatory requirement for new entities. Often, the regulatory agencies will audit the analytical laboratories of applicants, and transgressions from cGMPs and GLPs will be noted in 483s, as a reason for not approving a new product. Thus, an analytical laboratory should make sure that all the guidelines and regulations are applied precisely for the generation of data.

Competition in the industry is fierce, and confidentiality requires that members of the analytical laboratory do not leak detrimental information to the outside world. Publications and presentations of papers should be cleared by senior management with the help of the legal department.

The manager of the laboratory should emphasize that scanning of the environment is also important to determine new methodologies, technologies, and procedures and can affect the product lines of current products and new products under development. Scanning of the external environment also includes participating in scientific societies' meetings, and in trade associations' meetings, all designed to influence the recommendations of these associations for the benefit of science and of your products.

9.2.8 Ethical Behavior

Fraud in the development of analytical data, either deliberately or by accident, cannot be tolerated. The presence of a code of conduct is only the first step in ensuring ethical behavior from the members of the analytical laboratory. An ethical issue occurs when the behavior is not covered by the code of conduct but will affect the credibility of the organization data.

9.3 Management of Analytical Scientists

There are a number of issues that need to be discussed in the management of scientists and associated personnel in an analytical laboratory. These include technical issues, administrative issues, and managerial issues. Keep in mind the various management principles discussed under Section 9.2 [5].

9.3.1 Technical Issues Impacting the Management of an Analytical Laboratory

A number of chapters in this book detail the technical issues that could be encountered from a technical point of view. In this section, we look at these issues as they could affect the management of the laboratory [3]

9.3.1.1 Selection of Analytical Methods

The methods selected for a given product should be accurate and precise, rugged, reliable, meaningful, and consistent with the best practices in the industrial, governmental, and academia laboratories.

The selection of analytical methods will be affected by regulatory requirements, USP monographs, and the nature of the analysis, including timeliness of the results and economical considerations. This requires the laboratory management to decide what tests will be used, keeping in mind the various factors indicated earlier.

9.3.1.2 All Selected Methods Should Be Validated for Their Intended Purposes

The decision by the management of the laboratory to use a compendial method is a wise one, since by definition, the USP monographs' tests are validated by definition, and they are acceptable by FDA. However, you have to verify that the test is appropriate for a given product that does not interfere with the validated test [9].

Other methods can be used and are acceptable to FDA, if the laboratory develops data indicating that the new method is as good as or better than the compendial test. Reasons to use these methods are to take advantage of the newly developed procedures, new equipment, and automation. These considerations are important, especially from an economical point of view, since, for example, automation will increase throughput.

9.3.1.3 The International Congress on Harmonization (ICH) Factor

The harmonization of testing methods and requirements, brought about by The International Congress on Harmonization (ICH) [7] in combination with the pharmacopeias' initiative on harmonization [10], has had a significant effect on the work of analytical laboratories. Instead of testing products based on regional location, analytical laboratories can now test products using harmonized procedures, reducing the overall cost of testing.

9.3.1.4 Management of Analytical Laboratory and cGMPs and GLPs

An analytical laboratory has several functions. First, it will service QA and QC laboratories in the testing of current products, and management must ensure that cGMPs are followed. Second, the analytical laboratory service R&D develops analytical methods or validates the current methods for application to the new products, and that is done under GLPs. cGMPs and GLPs have different focuses and different requirements. It is up to the management of the laboratory to ensure that cGMPs and GLPs are followed, when appropriate.

9.3.1.5 Management under International Standardization Organization Certification

When inspecting analytical laboratories, FDA pays attention to the presence or absence of quality systems [8]. One of the aspects of the certification is the presence of appropriate SOPs. These ensure that performance of the analytical methods is consistent and that the results are properly documented. SOPs should be reviewed at appropriate interval or when some changes are introduced in manufacturing.

9.3.2 Administrative Issues

These issues include performance plans and performance appraisal of the scientists and support personnel, promotional criteria, training, retention, and hiring and firing of personnel.

9.3.2.1 Performance Plans and Appraisals

Few, if any, laboratory personnel are subjected to the development of performance plans, and performance appraisal always comes as a surprise to the personnel. In order to level the playing field and provide performance feedback, it is important for the manager of the laboratory to develop objectives and plans for each of the members of the laboratory. To obtain a buy-in from the personnel, you would want to develop these plans and objectives with the employees. The plans have to be doable but also challenging. Watch for the tendency from some scientists to dilute the objectives, in order to ensure that their performance will be easily attainable and not too challenging [2].

The objectivity of the performance appraisal is that the goals of the plans are compared to the actual accomplishments. Human resources in an organization will develop guidelines for the development of plans and for the appraisal, including the various timetables. However, the scientists and associated personnel dread these performance appraisals since they are generally tied to yearly merit increases. The manager of the laboratory does also feel very uncomfortable with these performance appraisals since they might involve confrontation with the personnel. A new approach to performance appraisals, including appraisal of performance by peers, has been tried and found lacking, since it has a tendency to introduce personal biases in the appraisals.

9.3.2.2 Training of Personnel and Promotional Opportunities

It is the responsibility of the manager of the laboratory to ensure that the personnel obtain appropriate training that would enrich the working experience and could be a factor in future promotional opportunities. The training subjects and schedules will have to be discussed with the personnel on an individual basis and privately. Sometimes, training might be induced because of poor performance on the job, and this should be separated from training designed to enrich the working experience.

Training can be integrated to the needs of the analytical laboratory and the areas of interest of the trainee. Since not everyone can be sent for training, due to a lack of funds, the manager can ask the trainee to give a seminar on the training received to other personnel, so that such training becomes a multiplier.

Promotional opportunities should also be planned for personnel that show by their performance that they are ready for additional responsibilities. Criteria or promotion at all levels should be developed and published, because transparency will enhance open communication established between the management and the working personnel. The retention of personnel is another factor that should be taken into consideration when training is selected, and promotional opportunities are envisioned.

9.3.2.3 Hiring and Firing of Personnel

Hiring in an analytical laboratory is designed to enhance the skill sets of the laboratory, especially when new and advanced developments in technologies and equipment have emerged. Technical knowledge and experience might be a prime consideration in hiring, but also potential fit within the laboratory should be taken into consideration. Referral from current employees is a way to reduce the uncertainty of hiring. It is also important that peers are also part of the interviews since the potential hire might feel more comfortable in being interviewed by peers rather than managers.

Firing of personnel must also be planned in detail with managers and supervisors seeking the advice of human resources. The track record of non-performance must be carefully documented and due process must be followed for legal and liability reasons

9.3.3 Managerial Issues in an Analytical Laboratory

A manager of an analytical laboratory will have to provide managerial input to planning, organizing, monitoring, and control of activities. He/she is also responsible for the productivity of the laboratory, budget development, and conflict management. Other inputs are also required for the outsourcing of tasks.

9.3.3.1 Planning

The manager of the analytical laboratory has to systematically analyze the activities being conducted, the activities that will need to be conducted, and the resources available to cover the current and near-future activities [4]. Any activity that does not contribute to the accomplishments of the organizational objectives must be justified. If not, they are candidates for removal.

Since "system thinking" should be operational, the objectives of the laboratory, however, should defer from the objectives of the organization that are prime.

Using a zero-budgeting approach, one can quickly and systematically determine if some activities are needed or not. Perhaps, an additional approach to planning is to also determine if old activities can be removed or reduced in scope or magnitude. The concept of the "Experience Curve" can also be invoked, since in routine testing, there is a correlation between the number of tests performed and the cost or duration of each test. Few, if any, laboratories do use the experience curve to reduce the cost of testing [2].

9.3.3.2 Organizing

Organizational structure does have its bearing on the activities of an analytical laboratory [3]. However, in most organizations, there are legacy structures that might hamper the development of a laboratory structure that is optimal for that organization. The generally accepted organizational structure is called a functional structure, where the analytical laboratory is divided into areas of expertise, such as a microbiology laboratory, a physical measurement laboratory, and a chemistry laboratory. The concentration of expertise within each of the functions is advantageous since the manager can assign specific tasks to each subfunction and ensure that the work will be done properly. A disadvantage of such a structure is that of "silo thinking," where newer solutions are difficult to implement and no technology transfer can easily be optimal. Another structure is the matrix structure, which for every project, such as the development of a new product, representatives from each of the subfunctions are brought together on a temporary basis to see that the needs of the projects are fulfilled. The advantage of such an organizational structure is that it promotes technology transfer, communication between the various subsections is enhanced, and perhaps, it can be translated into an increase of the probability of success of the project.

9.3.3.3 Monitoring and Control

Another management responsibility includes monitoring and controlling the performance of the activities [1]. The activities are compared to the plan in terms of duration, utilization of resources, quality of the output, and this should be done in a systematic way. Monitoring is done on a continuous basis and will trigger corrective actions if necessary. The manager of the laboratory controls the performance of the laboratory via monitoring and the implementation of corrective actions. These corrective actions should be contemplated when significant deviations occur due to events under the control of the manager or events not under the control of the manager of the laboratory.

When significant deviations occur, the manager should start an investigation to determine the impact of the deviations on the work of the laboratory, in terms of reduction in quality, reduction in output, or a combination of both. If the problems are not properly defined, the probability of a corrective action to be effective will not be very high. Rushing to implement corrective actions without defining the problem is a mode of operation that is frequently used in laboratories.

9.3.3.4 Resolution of Conflicts

Conflicts within an analytical laboratory exist and need to be managed. Conflicts can arise because of differences of opinions among the members of the laboratory, different personalities, management pressure, burnout, style of the manager of the laboratory, and even political issues.

If a conflict is low grade, the manager might not take any corrective action, since, in general, conflict at a low level might enhance the innovation and creativity in the resolution of technical problems. When a conflict, regardless of its origin, has the potential to affect the orderly function of the laboratory, the manager will have to intervene. Intervention will be on a one-to-one basis and not involve all the staff. Methods to reduce or minimize conflicts include confrontation between the conflict parties (preferred method), compromise (least preferred), smoothing, or withdrawal (not long-term solutions, since the conflict will reappear).

9.4 Conclusions and Recommendations

The integration of management in an analytical laboratory with the technical challenges in that laboratory will provide an organization in the pharmaceutical/biotech industry with a competitive edge that senior management should exploit. The "system thinking" principle indicates that the performance of each function of the organization should be directed toward the accomplishment of the organizational objectives. This is true for the analytical laboratory that occupies a central position in ensuring that products conform to specifications, regulatory and pharmacopeial requirements that data generated is credible and scientifically derived and can sustain the scrutiny of FDA, and that they are all designed to ensure the safety and effectiveness of the organization products for patients.

The crucial point in this chapter is that management principles must be applied if the analytical laboratory's desire is to optimize its contributions to the organization as a whole. It is thus necessary that the manager of the analytical laboratory be versed in these management principles, or if he/she is not, that appropriate training be provided. Overlaying all the management principles described in this chapter is that communication skill sets are necessary. The manager should utilize the open communication channels to transmit to the scientists the importance of their performance. Often neglected is communication with associated personnel (technicians, administrative personnel) without which a laboratory might falter.

Abbreviations

ANSI	American National Standard Institute
cGMPs	current good manufacturing practices
GLP	good laboratory practice
ISO	International Standardization Organization
FDA	Food and Drug Administration
QA	quality assurance
QC	quality control
NF	National Formulary
SOP	standard operating procedures
USP	United States Pharmacopeia

References

1 Dabbah, R. Total Project Management: Strategies and Tactics for the Healthcare Industry. Buffalo Grove, IL: Interpharm Press, (1993).

2 Dabbah, R.. Total R & D Management: Strategies and Tactics for 21st Century Healthcare Manufacturers. Buffalo Grove, IL: Interpharm Press, (1999)

3 Dabbah, R. Management issues in the microbiology laboratory in Laboratory Design-Establishing the Facility and Management Structure, S. Sutton, ed., Bethesda, MD & River Grove, ILL: PDA and DHI Publishing, (2010), pp. 15–43

4 Dabbah, R. Management of the microbiology function of organizations in Microbiology in Pharmaceutical Manufacturing, R. Prince ed., Bethesda, MD & River Grove, IL: PDA and DHI Publishing, (2001), pp. 759–778

5 Drucker, P. *Tasks, Responsibilities, and Practices.* NY: Harper & Row, (1974), p. 67

6 Federal Food, Drug, and Cosmetic, 21 USC, Chapter V, 2.7, Jan. 2001.

7 International Congress for Harmonization 2015. *The ICH Process*, www.ich.org/about/process-of-harmonization.html.

8 International Standardization Organization (ISO), Quality Systems and Quality Management, ISO 9000–9004. American Society of Quality Control (ASQC)/American National Standard Institute (ANSI), most current editions.

9 <1226> Verification of Compendial Procedures, USP 39-NF 34, United States Pharmacopeial Convention, Rockville, MD.

10 USP Pharmacopeial Harmonization<1196>, USP 36-NF 31, United States Pharmacopeial Convention, Rockville, MD.

11 Westland, J. What are Soft Skills? Plus. Free Assessment Template!, Project Management 101. https://www.projectmanager.com/blog/what-are-soft-skills-plus-free-assessment-template, (2016).

10

Analytical Instrument Qualification

Shaligram Rane and Rustom Mody

10.1 Introduction

Qualification is the process of ensuring that an instrument is suitable for its intended application.

Analytical results are most important in the pharmaceutical industry, and based on the analytical results, product quality at each stage decides the next course of action, whether it is raw material, in-process material, intermediate, drug substance, or drug product. In broader terms, the analytical results decide product quality in terms of a drug's safety and efficacy. From the good manufacturing practice (GMP) perspective, product quality should be built in the process by way of consistently delivering the expected quality attributes, which in turn depend on the quality of analysis. Quality of the analytical results depends on various factors. Among these, the most important ones are the analyst, instrument, and method.

Nowadays, sophisticated analytical equipment are available with state-of-the art optics, detectors, accuracy, speed, automation, statistical data analytical tools, and software. During the selection of analytical equipment, it is important to know the purpose or the analytical role, which should be well written in advance. This is referred to as the "User Requirement Specifications." In most of the pharmaceutical industry, user requirement specifications are made as part of a regulatory compliance requirement rather than based on the scientific rationale behind the objective of analysis and quality being investigated, defeating the very purpose of instrument selection. Those who are really knowledgeable in the respective field should write the user requirement specifications keeping in mind the technical capabilities that the instrument offers, the instrument's sensitivity and precision, orthogonal determination of quality attribute, robustness of analysis, and so on. The ultimate purpose is to generate reliable test results based on scientific principles.

The choice of an analytical instrument is often based on subjective likes and dislikes based on ease of operation, simplicity, data analysis, speed, and so on,

Pharmaceutical Analysis for Small Molecules, First Edition. Edited by Behnam Davani.

rather than on the basis of good scientific principle, sensitivity, robustness, precision, robustness, and statistical capabilities. The latter is what the regulators expect.

In this chapter, we propose an effective decision-making based on science- and risk-based approach for laboratory instrument selection. The chapter also covers analytical instrument qualification process from identifying the right quality of analytical instrument to its routine usage including periodic verification, calibration, and maintenance of equipment throughout its shelf life.

The main goal in qualifying the laboratory equipment is to ensure the validity of data. This calls for a robust instrument qualification program. Qualification shall involve science-based approach to provide documented evidence that the instrument is capable of consistently operating within established limits and tolerances.

The current equipment qualification procedures used within the pharmaceutical industry are designed on the basis of regulatory expectations and vendor recommendations. Equipment qualification plan must be well defined and documented. The plan shall have detailed procedure for the qualification of all laboratory equipment used for collection of data for release of intermediates, active pharmaceutical ingredient, formulated drug, stability testing, process or product characterization, and regulatory submissions.

For the qualification/validation program, the key element is validation master plan (VMP). It should be concise and clear and contain at least of the following:

- A validation policy
- Organizational structure of validation/qualification activities
- Roles and responsibilities of staff responsible for qualification/validation
- Summary of facilities, systems, equipment, and processes validated and to be validated
- Documentation format (e.g., protocol and report format)
- LIVE validation status of each equipment, process, method
- Planning and scheduling, and so on

10.2 Definitions

- **Audit Trail:** A secure, computer-generated, time-stamped electronic record that allows reconstruction of the course of events relating to the creation, modification, and deletion of an electronic record including the reason for the change. It must be ensured that any changes to the electronic data do not obscure the original data or earlier modifications and that there is no possibility to alter the audit trail of changes made to an electronic record.

- **Commissioning:** An engineering term that covers all aspects of bringing a system or subsystem to a position where it is regarded as being ready for use in pharmaceutical manufacture. Commissioning involves all the basic requirements before the start of installation qualification (IQ) and operational qualification (OQ).
- **Certified Copy:** A copy of original information that has been verified, as indicated by the dated signature, as an exact copy having all of the same attributes and information as the original.
- **Computer System (CS):** A group of computer (hardware), software, and associated documents (e.g., user manual) designed and assembled to perform a specific function or a group of functions such as to create, modify, maintain, archive, retrieve, or transmit information in digital form.
- **Design Qualification (DQ):** Documented verification that the proposed design of equipment or system is suitable for the intended purpose.
- **Electronic Record:** Means any combination of text, graphics, data, audio, pictorial, or any other information represented in digital form that is created, modified, maintained, archived, retrieved, or distributed by a computer system.
- **Electronic Signature:** Means a computer data compilation of any symbol or a series of symbols, executed, adopted, or authorized by an individual or combined identification codes/passwords or electronic signatures at the start of a data entry session to be the legally binding equivalent of the individual's handwritten signature.
- **Factory Acceptance Test (FAT):** Inspection and static and/or dynamic testing of system or major system components to support the qualification of an equipment system conducted and documented at the supplier site.
- **Impact Assessment:** The process of evaluating the impact of a system on the process and product quality. This assessment is done to formulate the qualification strategy for the system, based on its impact on the process and/or product, and to identify the critical components within those systems.
- **Installation Qualification (IQ):** Documented verification that the equipment or systems, as installed or modified, comply with the approved design, the manufacturer's recommendations and/or user requirements.
- **Operational Qualification (OQ):** Documented verification that the equipment or systems, as installed or modified, perform as intended throughout the anticipated operating ranges.
- **Performance Qualification (PQ):** Documented verification that the equipment and ancillary systems, as connected together, can perform effectively and reproducibly based on the approved process method and specifications.
- **Performance Verification (PV):** Periodic documented verification that the equipment and ancillary systems, as connected together, can perform effectively and reproducibly based on the approved process method and specifications.

- **Piping and Instrument Diagrams (P&IDs):** Engineering schematic drawings that provide details of the interrelationship of equipment, services, material flows, plant controls, and alarms. The P&IDs also provide the reference for each tag or label used for identification.
- **Qualification:** Identification of equipment attributes related to the performance of a particular function or functions and allocation of certain limits or restrictions to those attributes. Action of proving and documenting that equipment or ancillary systems are installed properly, operate correctly, and as expected, actually lead to the expected results.
- **Risk Assessment:** Method to assess and characterize the critical parameters in the functionality of an equipment or process.
- **System:** A group of equipment with a common purpose.
- **Security:** Preventing unauthorized access to data or records, accidental or intentional data manipulation, corruption or destruction of data through factors such as power failure or computer virus.
- **Software Validation:** Means confirmation by examination and provision of objective evidence that software specifications conform to user needs and intended uses, and the particular requirements implemented through the software can be consistently fulfilled before the software is delivered to the end user. Commercial software used in electronic recordkeeping systems needs to be validated as per the end-user requirement specification.
- **Transmit:** Means to transfer data from the site of generation to a remote site within the facility through network/cable for the purpose of retrieval, printing, reconstruction, analysis, or storage.
- **User Requirement Specification (URS):** User requirement specification is the document prepared by a user that enlists the requirements of the equipment or systems, before its procurement.
- **Worst Case:** A condition or a set of conditions encompassing upper and lower processing limits and circumstances, within standard operating procedures, which pose the greatest chance of product or process failure when compared to ideal conditions. Such conditions do not necessarily induce product or process failure.

10.3 Qualification: General Flow

See Figure 10.1.

10.4 Qualification Strategy: V Model

See Figure 10.2.

Figure 10.1 General concept of qualification activity.

10.5 Qualification

10.5.1 Qualification Scheme for New Equipment

Qualification phase provides documentation that all the critical equipment and utilities are installed properly through an installation qualification (IQ) and are operated correctly through an operational qualification (OQ), as well as that they perform effectively through a performance qualification (PQ). Qualification assures that on the criteria set forth, the basis of design is met. The procedure may be applied to individual equipment or a system (a group of equipment that together serve a common purpose) and shall apply to all new

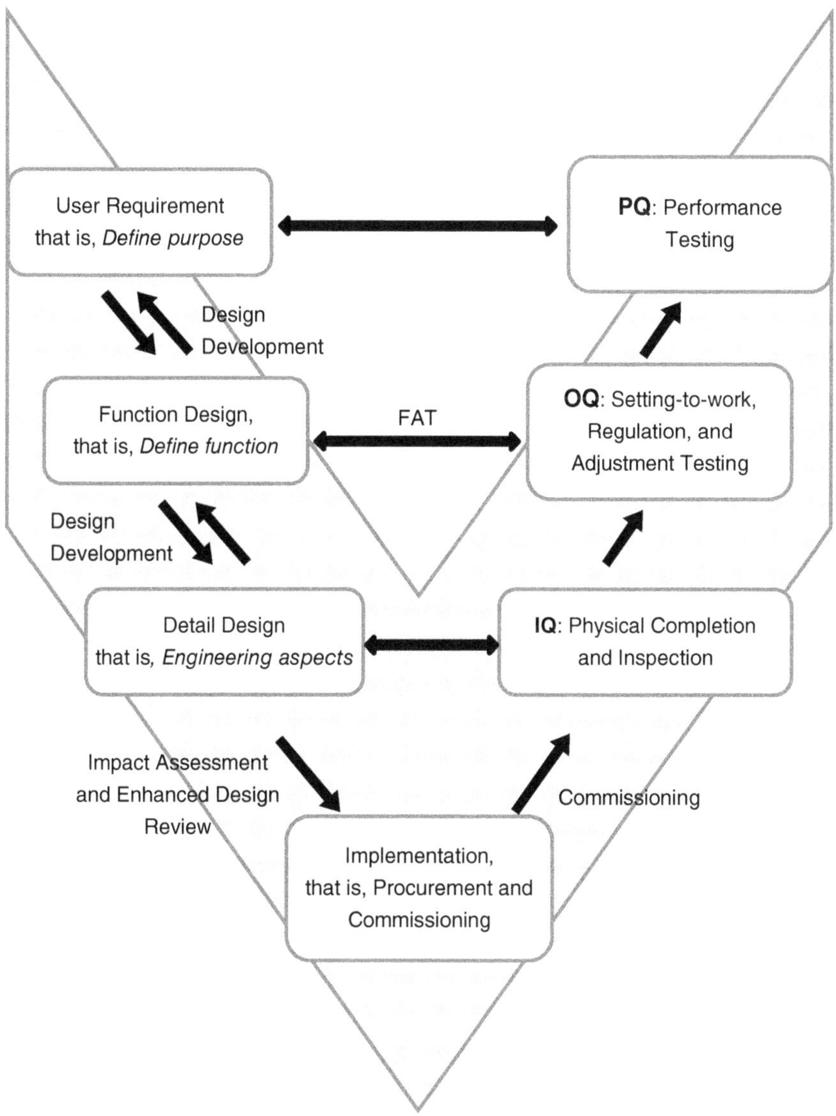

Figure 10.2 V model for the execution of qualification activity including flow of documentation.

as well as refurbished or used equipment being procured or transferred from one location to another or one application to another within the company.

Impact assessment can be categorized into direct impact, indirect impact, and no impact.

- Direct impact systems are subjected to qualification activities that incorporate design review, control, and testing against specification or other requirement necessary for cGMP compliance. All equipment identified as direct impact during impact assessment shall undergo qualification activities described as follows, starting with URS to PQ.
- Indirect impact systems are designed and commissioned following good engineering practices only.

Instrument Category:
Based on the complexity of the instrument being used in modern laboratories, USP general chapter <1058> recommends categorizing it by considering the criticality of the usage of each equipment. It is summarized in Table 10.1

Vendor Qualification: Generally, laboratory equipment are not customized, but for customized critical equipment, vendor qualification shall be carried out

Table 10.1 Recommended group based on complexity of instrument.

Group	Description	Requirements	Examples
A	Standard equipment with no measurement capacity	Verified and documented by visual observation	Nitrogen evaporators, magnetic stirrers, vortex mixers, centrifuges
B	Instruments and equipment providing measured values; Equipment controlling the physical parameters (temperature, pressure, flow) that need calibration	Conformance determined by SOPs and documented during IQ and OQ	Instruments: balances, melting point apparatus, light microscopes, pH meters, variable pipets, refractometers, thermometers, titrators, viscometers Equipment: muffle furnaces, ovens, refrigerators, freezers, water baths, pumps, and dilutors
C	Instruments and computerized analytical systems	Performance (OQ and PQ) limits are specific for analytical application: require specific function and performance tests. Full qualification process (may require assistance of a specialist)	Atomic absorbance spectrometers, differential scanning calorimeters, dissolution apparatus, electron microscope, flame absorption spectrometers, HPLCs, MS, microplate readers, thermal gravimetric analyzers, X-ray fluorescence spectrometers, X-ray powder diffractometers, densitometers, diode-array detectors, elemental analyzers, gas chromatographs, IR spectrometers, near-IR spectrometers, Raman spectrometers, UV/Vis spectrometers, inductively coupled plasma-emission spectrometers

before the design qualification phase. Vendor qualification for off-the-shelf, standard equipment with limited customization may not be carried out. Based on good reputation, or past experience, good reference from other users, ISO 9001 certification, or similar accreditations shall be considered sufficient.

10.6 Qualification Phases

10.6.1 User Requirement Specification

User requirement referred for selection of the equipment. For all new equipment, systems, support systems, utilities, URS shall be prepared. URS shall have detailed information regarding the following:

- Technical requirement
- Functional requirement
- Material of construction
- Utility requirement
- Safety feature requirement
- Interlocks requirement
- Alarm requirement
- Printing requirement
- Data storage or acquisition requirement
- Supporting documentation/certificates requirement, and so on

The URS should be a point of reference throughout the validation life cycle.

URS should be prepared by competent QC person, reviewed by QA, Engineering, IT person and approved by QA person. During finalization of URS, following parameters shall be considered:

Technical requirements – parameters, specification, and so on

Utility requirement, safety requirement, and IT requirement and documents required from vendors

10.6.2 Impact Assessment

Before inception of formal qualification, impact assessment is good to decide the criticality of equipment/instrument, which will help to decide the qualification strategy and depth of qualification.

In case of certain instrument, that is, less critical instruments, IQ and OQ can be merged as IOQ, while OQ and PQ can be merged as OPQ. In few cases only, commissioning will suffice requirements.

During assessment, impact can be categorized into three levels:

Level-1: Direct impact

Level-2: Indirect impact

Level-3: No impact

Level-1: Direct Impact
Any equipment or system that is part of a validated manufacturing process will normally be a direct-impact system.

- Direct impact on product quality.
- Assess the boundaries between direct and indirect impact systems.
- Subject to qualification.

For example: HPLC, GC, UV/Vis spectrometers, pH meter, autotitrator, and so on.

Level-2: Indirect Impact
No direct impact on product quality, but typically will support a direct-impact system.

- Designed and commissioned following good engineering practice only.
- Can affect the performance or operation of a direct impact system. Interfaces need to be carefully assessed.

For example: vortex mixture, stirrer, water bath

Level-3: No Impact
No direct impact on product quality, but typically will support some impact if specific parameter is required during analysis of the sample.

- Designed and commissioned following good engineering practice only.
- Can affect the performance or operation of a direct-impact system. Interfaces need to be carefully assessed.

For example: Vacuum pump

10.6.3 Design Qualification

The next element in the qualification of equipment, facilities, utilities, or systems is DQ, where the compliance of the design with GMP should be demonstrated and documented. The requirements of the user requirements specification should be verified during the design qualification.

Based on URS, equipment manufacturer/vendor shall prepare the design qualification (DQ), which should be reviewed and agreed by the user. The DQ shall contain all the technical details and design/process aspects of the equipment/instrument against each requirement mentioned in URS. For complex equipment, detailed layout, that is, P&ID, shall be a part of DQ. Reference URS number shall be a part of DQ.

DQ can be skipped in case the instrument is a catalog model/standard model. In case of analytical instrument, almost all instruments are catalog models/standard models available in the market.

For any old/existing equipment that is purchased earlier, if DQ is not available, based on its usage and criticality, DQ can be prepared and approved on the current date.

The DQ activity is most suitably performed by the instrument developer/manufacturer. Since the instrument design is already in place for the commercial off-the-shelf (COTS) systems, the user does not need to repeat all aspects of DQ. However, users should ensure that COTS instruments are suitable for their intended applications and that the manufacturer has adopted a quality system for developing, manufacturing, and testing. Users should also establish that manufacturers and vendors adequately support installation, service, and training. Methods for ascertaining the manufacturer's design qualification and an instrument's suitability for its intended use depend on the nature of the instrument, the complexity of the proposed application, and the extent of users' previous interaction with the manufacturer, vendor audits, availability of vendor-supplied documentation and certification, which satisfies the DQ requirement.

The required scope and comprehensiveness of the audits and documentation vary with users' familiarity with the instrument and their previous interactions with the vendor. Informal personal communications and networking with peers at technical or user group meetings significantly inform users about the suitability of instrument design for various applications and the quality of vendor support services. Informal site visits to other user and/or vendor facilities to obtain data on representative samples using the specified instruments are also a good source of information regarding the suitability of the instrument design for the intended use. In many instances, an assessment of the quality of vendor support, collected from informal discussions with peer users and vendor's technical service department, significantly influences instrument selection.

10.6.4 Factory Acceptance Test (FAT)

Equipment, especially if complex, may be evaluated, if possible, at the vendor site prior to delivery. Prior to installation, equipment should be confirmed to comply with the URS/functional specification as per user's requirement at the vendor's site.

Documentation review and some tests on the equipment can be performed during FAT. Repetition of such tests during IQ/OQ can be skipped if it can be shown that the functionality is not affected by the transport and installation.

10.6.5 Site Acceptance Test (SAT)

Once equipment arrives at the user site, site verification (also known as site acceptance test) may be executed to confirm that no physical damage to the equipment during shipping or transportation has occurred.

Testing at site (SAT) would remain somewhat similar to that performed during FAT.

10.6.6 Installation Qualification (IQ)

Once the equipment arrives, the installation phase begins. IQ shall be referred to as STATIC approach as it includes checkpoints by keeping the equipment in static mode.

Installation qualification should provide documented evidence that the installation was complete and satisfactory.

The first part of the qualification cycle requires formal documentation against installation checkpoints, that is, purchase specifications, drawings, manuals, spare parts lists, and vendor details.

It includes verification of the correct installation of components, instrumentation, equipment, pipe work, and services against the engineering drawings and specifications.

Control and measuring devices should be calibrated at this point.

In addition to the documentation of the program and its associated procedures, equipment shall be tagged or labeled and have records maintained. The labeling is for equipment identification.

Reference DQ/FAT number shall be a part of IQ. Logbook shall also be assigned to each equipment, and all details such as failures, maintenance, qualification testing, location, custodian, calibration records, servicing shall be recorded.

The objective of this installation qualification (IQ) protocol is to verify that the installation is done in accordance with the design and user requirements and meets the set acceptance criteria and cGMP requirements as stipulated in this protocol.

To verify that the requirements specified at the time of purchase are met in the delivered and installed item. Purchase order and equipment specifications have been used to prepare this protocol. Confirmation of the installed system to the predetermined specifications will verify that user requirements have been met.

Following points shall be considered during installation:

System identification – Equipment name, model number, equipment serial number, capacity, PO number, equipment manufacturer and address, vendor contact details

Documentation requirement – After execution, executed IQ protocol, all printouts, and handouts generated during installation, any laboratory test results record or their reference, any changes or deviation.

List of certificates

List of major components

List of safety measures and alarms

10.6.7 Operational Qualification (OQ)

Operational qualification should provide documented evidence that utilities, systems, or equipment and all its components operate in accordance with operational specifications.

IQ will be followed by OQ, which ensures that the equipment operates within the stipulated criteria as prespecified for each test.

OQ generally includes testing listed hereafter, but not limited to

- Operational checks
- Functional checks
- Programmable logic challenge test
- Alarms/interlock test,
- Printing verification
- Operational test
- Calibration
- Audit trail
- Access security and authentication controls
- Data analysis, storage, and retrieval.

Tests should be designed to demonstrate satisfactory operation over the normal operating range as well as at the limits of its operating conditions (including worst-case conditions).

Operation controls, alarms, switches, displays, and other operational components should be tested.

10.6.8 Performance Qualification (PQ)

PQ should normally follow the successful completion of IQ and OQ. However, it may in some cases be appropriate to perform it in conjunction with OQ or process validation

Performance qualification involves the testing of the equipment using the specific method or assay and material to ensure that the method is producing valid data.

PQ should provide documented evidence that utilities, systems, or equipment and all its components can consistently perform in accordance with the specifications under routine use.

PQ may consist of method validation testing, system suitability testing, analysis and trending of control samples. PQ testing procedures should be based on good science.

Performance qualification (PQ) is the documented collection of activities necessary to demonstrate that an instrument consistently performs according to the specifications.

10.6.9 Performance Verification (PV)

Once the equipment is qualified, its performance shall be verified on periodic basis. Some qualification test should be performed on a periodic basis, and this practice should be defined in the written qualification procedures.

Performance verification (PV) should be done in accordance with a defined protocol. PV testing may be the same as those verified during PQ. Based on criticality, the frequency of PV may vary.

10.6.10 Requalification

Requalification is applicable only in case of shifting or transfer of equipment from one location to another location, as well as requalification after changes such as changes to utilities, systems, equipment; maintenance work; or movement.

Requalification should be considered as part of the change control procedure.

During requalification, combined qualification approach can be adapted but needs to consider IQ, OQ, and PQ as good as new equipment. It varies with the type of equipment.

- Equipment/system may be subjected to requalification under the following criteria:
  - Change in location of system
  - Change in utilities
  - Any major modification or replacement of a critical component in the system
  - Change in intended use of the system
- Requalification of equipment shall be subjected to the "Impact Assessment" procedure to redefine the strategy for qualification.
- Extent of qualification and selection of the tests to be conducted for qualification/verification shall be decided based on risk assessment.

10.7 Qualification Issues

- Specifications not approved
- Failure to provide clear, complete instructions in the protocol
- Instrument calibration records or status not included
- Full range of intended operating parameters not challenged
- Inadequate sample sizes
- Unexplained deviations from protocol
- Inconsistencies between final report and data collection/recording forms
- How may runs are to be performed during operational qualification (OQ) testing?
- If one type and model of equipment is qualified, can it be used in a different process without additional qualification?

- What about "old manufacturers" who have not performed DQ or IQ for existing, in-use systems and/or equipment?
- What if requalification test fails?

10.8 Combined Qualification Approach/ Commissioning

In case of a certain instrument, that is, less critical instruments, IQ and OQ can be merged as IOQ, OQ and PQ can be merged as OPQ. Installation and operational checks, jointly, shall be a part of IOQ while operational and performance checks, jointly, shall be a part of OPQ.

In few of the cases only commissioning will also suffice qualification requirements.

10.9 Risk-Based Approach

It is good to have the quality risk management (QRM) program for equipment/ instrument qualification. Process flow is well described in "ICH Q9: Quality Risk Management," which includes steps such as *risk assessment, risk control, risk communication*, and *risk review*.

Once URS is finalized, QRM shall be carried out to find the risk, if any, during qualification and its control (reduction or mitigation) during qualification, that is, IQ/OQ/PQ. Based on QRM, one can decide which tests, at what frequency, shall be performed during performance verification.

10.10 Calibration/Verification

Calibration and verification of instruments and other devices, used in production and quality control, should be performed at regular intervals.

Personnel who carry out calibration should have appropriate qualifications and training.

A calibration program should be available and should provide information such as on calibration standards and limits, responsible persons, calibration intervals, records, and actions to be taken when problems are identified.

There should be traceability to standards (e.g., national, regional, or international standards) used in the calibration.

Calibrated instrument and other devices should be labeled, coded, or otherwise identified to indicate the status of calibration and the date on which recalibration is due.

When the instrument and other devices have not been used for a certain period of time, their function (requalification) and calibration status should be verified and shown to be satisfactory before use.

10.11 Track Performance Verification/Calibration Due Date

Tracking due date of performance verification or calibration is important to complete these tasks timely. One can prepare the tracking schedule for the entire year to avoid chances of frequency lapse.

10.12 Warning Letters Related to Laboratory Equipment

- *Inadequate laboratory equipment calibration program: failure to have written procedures describing specific calibration instructions and limits.*
- *Failure to conform to the USP section «41» for weight and balance determination.*
- *The inspection revealed that erroneous values are being used to perform the minimum weight studies. No certification to a recognized standard for the weights set used for checking the balance.*
- *The calibration procedure for HPLC systems is inadequate in that it did not include integrator's and detector's linearity, injector's reproducibility, and accuracy of temperature settings for column heater and detector.*
- *There are no predetermined acceptance criteria for the HPLC autosampler calibration.*
- *Procedures for UV/VIS spectrophotometer only assess linearity using alkaline potassium chromate solution at one wavelength when analytical tests are performed at various wavelengths. The procedures do not include functional tests such as wavelength accuracy, photometric accuracy, and reproducibility within ranges of intended use for the instrument.*
- *Calibration raw data and results obtained for the performance qualification of analytical instruments are not being checked for accuracy and completeness by a second analyst or laboratory supervisor.*

10.13 Equipment Qualification/Validation and Its Importance

The chapter intends to highlight the existing lacunas in the operational and performance qualification of HPLC instrument by relevant case study.

Routinely, the OQ and PQ of HPLC are carried out under the real laboratory conditions.

However, no challenge study is conducted to these "specific" laboratory conditions by varying the ambient temperature/airflow (importantly, for labs user-controlled air-conditioning units are installed). We have observed a wavy baseline if airflow from AC units is not uniform/consistent during the duration of analysis.

We also recommend considering HPLC placements in the laboratory.

Maintaining ambient temperature is vital for consistent performance of the HPLC instrument to avoid such baseline drifts and shift in RT of sensitive methods. Thus, during OQ, the room temperature should be varied, and the effect should be documented.

In routine practice, the HPLC qualification is performed with caffeine. The chromatogram of caffeine shows that the peak(s) are well defined with base-to-base separation. This enables precise and accurate results by autointegration, which leads to consistent OQ/PQ results over the lifetime of the instrument.

However, when such qualified systems are used to test biomolecules (recombinant proteins) with multiple impurities, the ranges set from OQ/PQ with chemical molecules such as caffeine become irrelevant. For biomolecules, PQ should be performed with the product of interest so as to achieve meaningful instrument operating ranges and method system suitability criteria.

These in-house functionally tests will be the zero-time data point for HPLC module. Subsequent PQ test, for example, after major repair/maintenance of the instrument, shall generate data for any drift over the life cycle of the instrument.

The following case study shows the importance of in-house testing during PQ.

The RP-HPLC test for protein content can get highly affected by the injector volume precision and accuracy.

As the total peak area is important to calculate the protein content, SST was primarily based on the injector precision. The system suitability criteria for injector precision were set with duplicate injections as 1.0%, which was equivalent to the manufacturer's specification (%RSD of area ≤1.00% with six injections of caffeine). However, protein binding and impurity resolution are different in biomolecules as compared to caffeine, which is comparatively a very pure form. (This could be because gradient accuracy and precision are not part of day-to-day PQ.)

This is depicted from the difference in the main peak areas of these molecules.

Thus, the PQ should be performed with in-house protein and methods prior to setting SST criteria for injection repeatability as well as resolution to avoid frequent SST failures during routine operations.

Table 10.2 Protein content result by changing the volume to verify the accuracy level.

Volume injected	Protein content (mg/mL)
19.6	11.17
20.0	11.40
20.4	11.62
%RSD	2.00

Consider a drug product specification of 90–110%. If the injection volume is 20 μL, the protein content variation due to injector accuracy can be 2.00% (Table 10.2).

Considering only injection accuracy factor, instrument variation of 2% can be deduced, which combined with column analyst variation might lead to a maximum method variation of greater than 5%. This combined with process variation may lead to OOS result.

Furthermore, if a stringent specification of 98–102% is required to be met, such method should have tighter SST criteria. Thus, a wise selection of instrument PQ tests should be done if such stringent specification criteria are required to be met.

The routine specifications of injector precision for different HPLC brands are as follows: %RSD for injector precision (Table 10.3).

Thus, functionality test of an instrument should include analysis of in-house methods using relevant test samples.

Table 10.3 Injector precision of two different HPLC brands.

Agilent	Dionex
2	0.3

10.14 Examples

10.14.1 HPLC (High-Performance Liquid Chromatography)

See Table 10.4.

10.14.2 UV/Visible Spectrophotometer

See Table 10.5.

Table 10.4 Parameters to be considered for HPLC during qualification and its importance [1–5].

Instrument module	Parameter to be checked	Tolerance limits	Importance of parameter to be checked	References
Solvent delivery system [1]	Flow rate [1]	±5% [1]	Consistent and accurate flow rate of the mobile phase is necessary to provide stable and repeatable interactions between the analyte and stationary phase. Important parameter to obtain the repeatable peak responses and elution pattern. In addition, important for comparability between systems and hence while transferring methods	[1, 2]
	Proportioning accuracy and precision (gradient test) [1]	Absolute deviation: ±2 [1]	This test is important to achieve the adequate chromatographic separation, consistent peak responses, elution pattern, and reproducibility. This ensures the ability of the pump/gradient proportioning valve (GPV) to deliver the mobile phase at different solvent strengths over the time by varying the composition of the mobile phase accurately	[1, 2]
	Proportioning ripple [1]	≤0.2% [1]	This test ensures the pressure observed during solvent mixing and thus ensures the proper functioning of pumps. This is usually measured in terms of percentage of noise of the 50% line from the gradient program	[1, 2]
Injector	Volume precision	RSD ≤1.0% [1]	This test ensures that the injector is able to withdraw the same volume of sample in replicate injections. This is crucial to precision and accuracy for the peak-area or peak-height comparison for the standards and samples. Measured by obtaining the repeatable peak areas, which in turn are measured in terms of % relative standard deviation	[1, 2]

(Continued)

Table 10.4 (Continued)

Instrument module	Parameter to be checked	Tolerance limits	Importance of parameter to be checked	References
	Volume linearity	$r^2 \geq 0.999$ [1]	HPLC injectors are capable of varying the injection volume without changing the injection loop. For example, A 100 μL loop can inject volume from 1 to 100 μL This variable volume of sample is drawn into the sample injection loop by a syringe. The uniformity of the injection loop and the ability of the syringe or other metering device to draw different volumes of sample in proper proportion will affect the linearity of the injection volume. Hence, linearity needs to be established for the injector, which ensures that the injector system works well for the defined volume ranges	[1, 2]
	Carryover	Blank I peak response: NMT 0.5% of reference solution (b) injection I Reference solution (c) peak response should not be more than ±10 % of reference solution (b) injection II [1]	Small amounts of analyte may get carried over from the sample injected before and lead to contamination of the next sample to be injected. This will affect the accuracy of the quantification of the next sample in queue. Hence, this test ensures the % of carryover from sample to sample. This is usually measured by injecting a high concentration of sample followed with the blank injections [2]	[1, 2]
Autosampler	Thermosetting accuracy and precision	±3 °C [1]	Protein samples are temperature-labile, and during analysis, appropriate temperature conditions also are required to be maintained. This test ensures the accuracy and precision of autosampler when the required temperature is set and compared with the actual values observed by using an external calibrated temperature probe [2]	[1, 2, 4]

Oven or cooling device	Thermosetting accuracy	±2 °C [1]	The ability to maintain an accurate column temperature is highly essential to achieve the desired retention time and resolution in the separation process. This test ensures the accuracy and precision of column compartment when the required temperature is set and compared with the actual values observed by using an external calibrated temperature probe [2]	[1, 2]
UV/DAD detector	Linearity	$r^2 \geq 0.999$ [1]	For the comparison of standards and samples, and accordingly the determination of analyte in these samples, it is very important for the linearity of the detector to be accurate for the peak area or peak height. This test ensures that when different concentrations of test solutions are injected over a desired range, then the response is linear, that is, the concentration is directly proportional to the observed main peak areas. In addition, it ensures that the vials are correctly detected by the equipment (vial positioning check) [2]	[1, 2, 4]
	Wavelength accuracy	±2 nm [1]	This test ensures that the detector accurately determines the absorption (maxima and minima) by using a well-characterized compound such as caffeine/anthracene/holmium oxide filter. This is usually measured as a deviation of the wavelength absorption from the known wavelength of band. This helps to ensure that the results obtained at a particular wavelength are well within the system acceptance criteria [2]	[1, 2, 4]

(Continued)

Table 10.4 (Continued)

Instrument module	Parameter to be checked	Tolerance limits	Importance of parameter to be checked	References
Fluorescence detector	Wavelength accuracy excitation	±3 nm	Lamp source with monochromator is used for the required wavelength. These gratings/monochromator may deteriorate over time, and therefore, the accuracy of the wavelength at which energy is absorbed or emitted needs to be verified	[1]
	Wavelength accuracy emission	±3 nm		
	Sensitivity	≤0.5 ppb	Sensitivity is the ability of the detector to determine the lowest amount of analyte	[1]
Electrochemical detector	Accuracy of cell current	±0.05 nA	The electrochemical sensor measures the electrons in the form of electrical energy, which is transferred due to electrochemical reaction. This sensor transfers the current to an electronic circuitry, which converts the pico- or nanoampere current in a signal in terms of volts. Hence, the results are measured as current/volt. Hence, it is important to test the accuracy of the cell	[1]

Table 10.5 Parameters to be considered for UV/visible spectrophotometer during qualification and its importance [6–9].

<table>
<tr><th>Parameter to be checked</th><th>Tolerance limits</th><th>Importance of the parameter to be checked</th><th>References</th></tr>
<tr><td>Control of wavelength (wavelength accuracy)</td><td><table><tr><th>Wavelength (nm)</th><th>Limit (nm)</th></tr><tr><td>241.15 (Ho)</td><td rowspan="8">±1</td></tr><tr><td>253.7 (Hg)</td></tr><tr><td>287.15 (Ho)</td></tr><tr><td>302.25 nm (Hg)</td></tr><tr><td>313.16 nm (Hg)</td></tr><tr><td>334.15 nm (Hg)</td></tr><tr><td>361.5 nm (Ho)</td></tr><tr><td>365.48 nm (Hg)</td></tr><tr><td>404.66 nm (Hg)</td><td rowspan="8">±3</td></tr><tr><td>435.83 nm (Hg)</td></tr><tr><td>486.0 nm (Dβ)</td></tr><tr><td>486.1 nm (Hβ)</td></tr><tr><td>536.3 nm (Ho)</td></tr><tr><td>546.07 nm (Hg)</td></tr><tr><td>576.96 nm (Hg)</td></tr><tr><td>579.07 nm (Hg)</td></tr></table>[8, 9]</td><td>Every compound has characteristic absorption maxima and minima wavelengths
Wavelength accuracy is the deviation of the wavelength reading at an absorption band or emission band from the known wavelength of band
Therefore, if the equipment is not able to maintain accurate wavelength scale, the UV absorption profile of the sample measured by the equipment may not be accurate, and the true λ max and λ min of an analyte cannot be determined appropriately
Example: most of the assays specify that absorbance of analyte to be considered at λ max. An analyte may have a broad absorption profile. Minor wavelength deviation from the λ max may not impact the results; however, if the selected wavelength is not close to the λ max, this will cause a major impact on the results. Significant wavelength deviation from the optimal wavelength of λ max also effectively reduces the extinction coefficient and thus the sensitivity of the measurement [6]</td><td>[6, 8, 9]</td></tr>
</table>

(Continued)

Table 10.5 (Continued)

<table>
<tr><th>Parameter to be checked</th><th>Tolerance limits</th><th>Importance of the parameter to be checked</th><th>References</th></tr>
<tr><td>Control of absorbance</td><td>
<table>
<tr><th>Wavelength (nm)</th><th>Solution used</th><th>A (1%, 1 cm)</th><th>Maximum tolerance</th></tr>
<tr><td>235</td><td rowspan="4">$K_2Cr_2O_7$ Solution (0.006 % w/v in 0.005 M H_2SO_4)</td><td>124.5</td><td>122.9 – 126.2</td></tr>
<tr><td>257</td><td>144.5</td><td>142.8 – 146.2</td></tr>
<tr><td>313</td><td>48.6</td><td>47.0 – 50.3</td></tr>
<tr><td>350</td><td>107.3</td><td>105.6 – 109.0</td></tr>
<tr><td>430</td><td>$K_2Cr_2O_7$ Solution (0.06 % w/v in 0.005 M H_2SO_4)</td><td>15.9</td><td>15.7 – 16.1</td></tr>
</table>
[8]</td><td>Photometric accuracy ensures the accuracy of the results obtained from the equipment by comparing the values from established standards with the known concentration of standards. Hence, this test should be performed at the final stage or at the last stage when all other test for spectrophotometer verification are performed, and this test provides higher degree of reliability for the results obtained [6]</td><td>[6, 8]</td></tr>
<tr><td>Limit of stray light</td><td>Absorbance should be greater than 2.0 at 198 nm [8]</td><td>Stray light is the light of any wavelength that is outside the bandwidth of the selected wavelength. The higher the amount of stray light, the absorbance will be less, and this may reduce the linear range of equipment. The linearity of the absorbance response is limited by the stray light at high absorbance. Therefore, absorbance values ranging from 0.3 to 1 are less susceptible to stray light and noise problems and hence become the preferred absorbance range for UV–visible analysis [6]</td><td>[6, 8]</td></tr>
</table>

Resolution	Should be greater than 1.5	Resolution determines if the two peaks are getting resolved when the analyte is scanned within a range of wavelength. The smaller the spectral band width, the finer the resolution. Spectral band width depends on the slit width and dispersive power of the monochromator	[6]
Cells	For 1 cm quartz cuvettes: absorbance should not be more than 0.093 at 240 nm For 1 cm glass cuvettes: absorbance should not be more than 0.035 at 650 nm Upon rotating the cells by 180°, the absorbance difference should not be greater than 0.005 units.	Continuous use of optical cells may lead to deterioration of cells. Hence, this test ensures cleanliness and gross differences in thickness or parallelism of windows of optical cells	[8]
Baseline flatness check	When air is scanned in the absorbance mode, the deflection in the highest and lowest absorbance is less than 0.01 AU [8]	The intensity of light sourced by tungsten and deuterium lamp varies over the entire wavelength range, and the detector response also varies. This test demonstrates the ability of the instrument to normalize the light intensity measurement by the detector and the spectral output at different wavelengths (light source: visible, tungsten lamp and UV region, deuterium lamp) throughout the spectral range [6]	[6, 8]
Spectral slit width (applicable only for equipment with variable slit width)	±10% of the selected spectral band width	Slit width should be selected appropriately to ensure receiving high value of light intensity but small enough compared to the half width of the absorption band. Hence, a slit width with variation of ±10% Does not result in a change in the absorbance reading	[8]

(Continued)

Table 10.5 (Continued)

<table>
<tr><th>Parameter to be checked</th><th>Tolerance limits</th><th>Importance of the parameter to be checked</th><th>References</th></tr>
<tr><td>Baseline noise</td><td>Mean ±0.002 AU (at 500 nm with no sample in the sample chamber)
<table><tr><th>Filter</th><th>Wavelength (nm)</th><th>Maximum tolerance</th></tr><tr><td rowspan="3">Quartz block (against air)</td><td>200</td><td>0.049 ± 0.01</td></tr><tr><td>300</td><td>0.033 ± 0.01</td></tr><tr><td>400</td><td>0.031 ± 0.01</td></tr></table>[8]</td><td>Noise in the UV–visible measurement originates primarily from the light source and electronic components. Noise affects the accuracy of measurements and thus makes the equipment less sensitive [6]</td><td>[6, 8]</td></tr>
<tr><td>Photometric drift/stability</td><td>±0.001 AU/h (time scan mode: 250 nm for 2 h)
±0.002 AU/h (time scan mode: 500 nm for 1 h) [8]</td><td>Lamp and electronics/optics may deteriorate over time, and using the instrument over extended period of time for the sample analysis may also result in drift/errors in the measurement. This test ensures the steady state of the equipment over a period of time and thus ensures that the effect of drift on the accuracy of measurements is insignificant [6]</td><td>[6, 8]</td></tr>
</table>

10.14.3 Autotitrator

See Table 10.6.

10.14.4 Karl Fischer Titrators

See Table 10.7.

10.14.5 Weighing Balance

See Table 10.8.

10.14.6 Auto Pipettes

See Table 10.9.

Table 10.6 Parameters to be considered for autotitrator during qualification and its importance [7, 10].

Parameter to be checked	Tolerance limits	Importance of the parameter to be checked	References
Potentiometric titrators			
Precision	RSD ≤0.2%	This test ensures that when the same standard/certified reference material is titrated minimum three or more times, and the % RSD complies as per the acceptance criteria, then the result/titer obtained as well as the method is precise. In addition, it ensures that the automatic burettes and aqueous/nonaqueous probes used are precise for the measurement	[10]
Accuracy	$d_{rel} \leq \pm 0.5\%$	This test ensures that the results obtained for the titer values are accurate. In addition, it ensures that the automatic burettes and aqueous/nonaqueous probes used are accurate for the measurement	[10]
Linearity	$r^2 \geq 0.9990$	When different weights of certified standard are used and titrated by using the automatic burettes and aqueous/nonaqueous probes, a linear straight-line curve with a regression value of greater than or equal to 0.9990 should be obtained, which ensures that the equipment gives linear results over the desired operating range	[10]

Table 10.7 Parameters of Karl Fischer titrators to be considered during qualification and its importance.

Karl Fisher volumetric titrators			References
Precision	RSD ≤ 1.0%	This test ensures that when the same standard/certified reference material is titrated minimum three or more times, and the % RSD complies as per the acceptance criteria, then the result/titer obtained as well as the method is precise. In addition, it ensures that the automatic burettes and F electrode used are precise for the measurement	[10]
Recovery	97.5–102.5%	This test ensures the accuracy of KF probe measurement, that is, the amount of water added for titration versus the value of water obtained by performing KF titration	[10]
Karl Fischer coulometric titrators			
Precision	Standard 1000 μg/g H_2O: RSD ≤ 2.0%	This test ensures that when the same standard/certified reference material is titrated minimum three or more times, and the % RSD complies as per the acceptance criteria, then the result/titer obtained as well as the method is precise. In addition, it ensures that the automatic burettes and KF coulometric probe used are precise for the measurement	[10]
	Standard 100 μg/g H_2O: RSD ≤ 5.0%		[10]
Recovery	Standard 1000 μg/g H_2O: 97.5–102.5%	This test ensures the accuracy of KF coulometric probe measurement, that is, the amount of water added for titration versus the value of water obtained by performing KF titration	[10]
	Standard 100 μg/g H_2O: 90.0–110.0%		[10]
Linearity	$r^2 \geq 0.990$	When different weights of certified standard are used and titrated by using the automatic burettes and KF coulometric probe, a linear straight-line curve with a regression value of greater than or equal to 0.9990 should be obtained, which ensures that the equipment gives linear results over the desired operating range	[10]

Karl Fischer oven used for microdetermination of water using an evaporation technique			
Temperature accuracy	±5.0 °C	KF titration measures the water content in the sample. In this case, evaporation technique is used for KF titration wherein the sample under the effect of heating releases water, which is carried through the carrier gas to the titration vessel; therefore, accurate temperature becomes important to be maintained to obtain accurate results	[10]
Temperature stability	±0.5 °C	As described under temperature accuracy, since evaporation technique is a bit time-consuming process to release water content and transfer it to the titration vessel, the stability of temperature is very important to be maintained to obtain accurate results	[10]
Carrier gas flow rate	±10 mL/min	Usually, air or nitrogen is used as the carrier gas for the transfer of released water content to titration vessel. The rate of transfer to the titration vessel also plays an important role in measurement readings	[10]
Precision	RSD ≤ 1.5%	This test ensures that when the same standard/certified reference material is titrated minimum three or more times, and the % RSD complies as per the acceptance criteria, then the result/titer obtained as well as the method is precise. In addition, it ensures that the automatic burettes and KF probe/oven/carrier gas supplied are precise for the measurement	[10]
Recovery	95.0–105.0%	This test ensures the accuracy of KF probe with oven, that is, the amount of water added for titration and the amount of water obtained by performing KF titration are accurate, and % RSD complies as per the acceptance criteria for recovery	[10]

Table 10.8 Parameters of weighting balance to be considered during qualification and its importance [7, 11].

Parameter to be checked	Tolerance limit	Importance of the parameter to be checked	References
Leveling	As per the acceptance limit of balance	Leveling ensures that the balance is stable from all four corners as well as the center, and the readings observed should become stabilized within the typical time required	[11]
Internal calibration (adjustment) Automatic/manual	Automatic as per the balance acceptance limits	Internal calibration/adjustment is a motor-driven comparison with in-built calibration weights. Internal calibration is more important when there are frequent condition changes in the area where balance is installed, such as temperature variation, power failure	[11]
Verification	As per the acceptance criteria of external NIST traceable weights used for verification	External verification with calibrated weights ensures that the displayed weight value, calibrated weight, and internal balance weights are in line with each other and thus ensures the accuracy of balances during daily usage	[11]
Accuracy	Acceptance criteria to be defined as per balance specifications	This test ensures that the displayed weight value, calibrated weight, and internal balance weights are in line with each other and thus ensures the accuracy of balances during daily usage	[11]

Linearity	R2: 1 ± 0.0001	Linearity of balance ensures that when weights of different capacities are placed on the weighing pan of balance, a linear response is observed between the theoretical and observed values	[11]
Precision	Maximum SD = 5**d* (*d* = actual scale interval, e.g., 0.1 mg)	Precision of balance ensures that when a weight of 50% of nominal balance capacity is measured for a minimum of five times, the balance should comply with the acceptance criteria, which demonstrates that the balance provides precise weight measurements	[11]
Eccentricity	RSD: NMT 0.05%	Eccentricity test is performed by weighing at least 30% weight of the maximum capacity of balance or the maximum weight to be used on balance onto different locations of the weighing pan to ensure that during routine usage also if the sample/material to be weighed is placed in any other location onto the weighing pan, the weight values obtained are well within the acceptance criteria	[11]
Linearity error	≤Accuracy of the balance	Linearity error is the difference between digital display for a weight that weighs 50% of full weighing balance capacity of the instrument and its true mass. This ensures that the balance gives linear results for different weights used	[11]

(Continued)

Table 10.8 (Continued)

Parameter to be checked	Tolerance limit	Importance of the parameter to be checked	References
Drift test (applicable only for semimicro, micro, and ultramicro balances)	NMT 0.05%	Drift is a progressive change in display (continuously upward or downward) of the digital readouts of balance, which means that weight readouts are not stable	[11]
Minimum weight	Based on external calibration certificate and technical data of balance	The minimum weight for which the balance can be used for weighing purpose with accuracy and precision	[11]
Measurement uncertainty	NMT 0.001%	Any of the variations due to the measurement variable can be measured under measurement uncertainty	[11]

Table 10.9 Parameters of auto pipettes to be considered during qualification and its importance [12, 13].

<table>
<tr><th>Parameters to be checked</th><th>Tolerance limit</th><th>Importance of parameter to be checked</th><th>References</th></tr>
<tr><td>Visual inspection</td><td>After 10 s, no drop is formed at the tip when tip is prewetted with a nominal volume of water</td><td>This test ensures that the auto pipette is working fine with no leakage observed</td><td>[12]</td></tr>
<tr><td>Gravimetric test for systematic error and random error</td><td>1. Type A: Fixed- and variable-volume monochannel pipettes (air displacement)
<table>
<tr><th>Nominal volume (μL) for variable volume</th><th>Maximum permissible systematic error (%)</th><th>Maximum permissible random error (%)</th></tr>
<tr><td>1</td><td>±5</td><td>±5</td></tr>
<tr><td>2</td><td>±4</td><td>±2</td></tr>
<tr><td>5</td><td>±2.5</td><td>±1.5</td></tr>
<tr><td>10</td><td>±1.2</td><td>±0.8</td></tr>
<tr><td>20</td><td>±1</td><td>±0.5</td></tr>
<tr><td>50</td><td>±1</td><td>±0.4</td></tr>
<tr><td>100</td><td>±0.8</td><td>±0.3</td></tr>
<tr><td>200</td><td>±0.8</td><td>±0.3</td></tr>
<tr><td>500</td><td>±0.8</td><td>±0.3</td></tr>
<tr><td>1,000</td><td>±0.8</td><td>±0.3</td></tr>
<tr><td>2,000</td><td>±0.8</td><td>±0.3</td></tr>
<tr><td>5,000</td><td>±0.8</td><td>±0.3</td></tr>
<tr><td>10,000</td><td>±0.6</td><td>±0.3</td></tr>
</table></td><td>Systematic error provides the accuracy of the auto pipettes, whereas random error provides the repeatability of auto pipettes
During routine usage of auto pipettes, there are chances of malfunction, and therefore at regular intervals, auto pipette calibration should be performed after every repair/maintenance
The gravimetric test ensures that the auto pipette is working with the required accuracy and precision over time</td><td>[12, 13]</td></tr>
</table>

(Continued)

Table 10.9 (Continued)

Parameters to be checked	Tolerance limit	Importance of parameter to be checked	References

2. **Type D**: Fixed- and variable-volume monochannel pipettes (positive displacement or direct displacement)

Nominal volume (μL)	Maximum permissible systematic error (%)	Maximum permissible random error (%)
5	±2.5	±1.5
10	±2	±1.0
20	±2	±0.8
50	±1.4	±0.6
100	±1.5	±0.6
200	±1.5	±0.4
500	±1.2	±0.4
1,000	±1.2	±0.4

3. **Type A**: Fixed- and variable-volume multichannel pipettes (air displacement)

Nominal volume (μL)	Maximum permissible systematic error (%)	Maximum permissible random error (%)
1	±10	±10
2	±8	±4
5	±5	±3
10	±2.4	±1.6
20	±2	±1.0
50	±2	±0.8
100	±1.6	±0.6
200	±1.6	±0.6
500	±1.6	±0.6
1,000	±1.6	±0.6
2,000	±1.6	±0.6
5,000	±1.6	±0.6
10,000	±1.2	±0.6

10.14.7 Gas Chromatography

See Table 10.10.

10.14.8 Analytical Column Qualification

See Table 10.11.

10.14.9 Melting Point

See Table 10.12.

Table 10.10 Parameters of gas chromatography to be considered during qualification and its importance [14].

Parameter to be checked	Tolerance limit	Importance of parameter to be checked	References
Injector leak test	Pressure drop ≤15 kPa within 5 min	This test ensures that there is no leakage from the injector when the column is disconnected and the injector outlet is closed, followed by applying pressure to the maximum for a period of 5 min	[14]
Headspace injectors: repeatability of headspace injectors)	Peak areas: RSD ≤ 5.0% Retention times: RSD ≤ 2.0%	When six consecutive injections of test sample are injected and peak areas, retention time comply as per the acceptance criteria, it ensures that the headspace injectors are precise to provide the results	[14]
Vial heater temperature	±4 °C from set point	Test solution needs to be maintained and injected at the appropriate desired temperature, and therefore, accuracy of the vial heater temperature is very important, which needs to be checked periodically by employing an external calibrated probe and comparing it with the displayed temperature	[14]

Table 10.11 Parameters of analytical column to be considered during qualification and its importance [15].

Parameter to be checked	Tolerance limit	Importance of the parameter to be checked	References
For RP-LC C-8 and C-18 columns (mixture of uracil, toluene, phenol, and *N,N*-diethyl *m*-toluamide is used)	$A_s = 0.8$ to 1.5 $R_s > 1.5$ (between adjacent peaks) The number of plates N for toluene >3000 Selectivity of α-phenol/toluene and α-*N,N*-diethyl-*m*-toluamide/toluene ≥1.5	Column qualification/verification can be either method-specific or as per the procedure and limits defined by the manufacturer. Asymmetric, resolution, theoretical plates, and selectivity of columns ensure that the column performance is well within the defined limits for further usage All these parameters provide information on column characteristics such as column strength, capacity	[15]
For RP-LC cyano columns (mixture of uracil, toluene, phenol, 4-cl nitrobenzene, and napthalene is used)	$A_s = 0.8–1.5$ The number of plates N for Naphthalene > 3000 $R_s \geq 1.5$ (between adjacent peaks)	Asymmetric, resolution, theoretical plates, and selectivity of columns ensure that the column performance is well within the defined limits for further usage All these parameters provide information on column characteristics such as column strength, capacity	[15]
RP-LC phenyl (phenyl propyl and phenyl hexyl columns) (mixture of uracil, acetophenone, toluene, and naphthalene is used)	$A_s = 0.8–1.5$ $R_s > 1.5$ (between adjacent peaks) The number of plates N for Naphthalene > 3000 Selectivity of α acetophenone/toluene and α toluene/naphthalene ≥ 1.5	Asymmetric, resolution, theoretical plates, and selectivity of columns ensure that the column performance is well within the defined limits for further usage All these parameters provide information on column characteristics such as column strength, capacity	[15]

NP-LC columns (mixture of toluene, diethyl phthalate, and dimethyl phthalate is used)	$A_s = 0.8–1.5$ $R_s > 1.5$ (between adjacent peaks) The number of plates $N > 3000$ Selectivity of α diethyl phthalate/dimethyl phthalate ≥ 1.5	Asymmetric, resolution, theoretical plates, and selectivity of columns ensure that the column performance is well within the defined limits for further usage All these parameters provide information on column characteristics such as column strength, capacity	[15]
Cation-exchange columns (mixture of uracil and cytosine are used)	$A_s = 0.8–1.5$ The number of plates $N > 3000$ $R_s > 1.5$ Selectivity of α uracil/cytosine ≥ 1.5		
Anion-exchange columns (mixture of uridine and uridine monophosphate are used)	$A_s = 0.8$ to 1.5 The number of plates $N > 3000$ $R_s > 1.5$ Selectivity of α uridine/uridine monophosphate ≥ 1.5		
Size-exclusion columns (gel filtration standard containing thyroglobulin, γ,γ-globulin, ovalbumin, myoglobin, and vitamin B12)	Plate number (vitamin B12): NLT 20,000 Symmetry factor (vitamin B12): 0.8–1.5 Resolution (myoglobin/ovalbumin): NLT 2.5		

Table 10.12 Parameters of melting point apparatus to be considered during qualification and its important [16, 17].

<table>
<tr><th>Parameter to be checked</th><th>Tolerance limits</th><th>Importance of the parameter to be checked</th><th>References</th></tr>
<tr><td>Clear point for three capillaries</td><td>±0.3 °C of each other [16]</td><td>This ensures that the heating block temperature is homogeneous/ symmetric over the complete surface</td><td>[16, 17]</td></tr>
<tr><td>Melting range of CRS</td><td>Less than 2 °C (at ramp rate of 1 °C/min) [16]</td><td rowspan="2">When different MP standards are used for calibration in the desired operating range, the range of melting point should be less than 2°. Every standard has a specific melting range, and any shift from that represents the failure in accuracy of the equipment</td><td></td></tr>
<tr><td>Melting point accuracy</td><td><table><tr><th>Melting point (°C)</th><th>Accuracy (°C)</th></tr><tr><td><100</td><td>±0.3</td></tr><tr><td>100–250</td><td>±0.5</td></tr><tr><td>>250</td><td>±0.8</td></tr></table>[16]</td><td></td></tr>
</table>

10.15 Qualification Status of Existing Equipment/ Instrument

This is the generic numbering system suggested. Numbering can be done as per individual company policy. This is just an example for reference, how to maintain the overall status of analytical instrument for quick reference (Table 10.13).

10.16 Summary

Analytical chemistry plays an important role in the pharmaceutical industry and totally depends on the setup, instrument quality, analyst, and so on. Selection of an analytical instrument is critical and depends on the intended application. Even after selection, qualification is an important aspect to ensure suitability of the instrument for its intended application. The decisions about product quality are made based on the analytical results. The ultimate purpose is to generate reliable test results based on scientific principles.

In this chapter, we proposed an effective decision-making based on science- and risk-based approach for laboratory instrument selection. The chapter also covers the analytical instrument qualification process from identifying the right quality of analytical instrument to its routine usage including periodic

Table 10.13 Overall status of the analytical instrument used for the testing of various samples in the laboratory.

			Qualification status (mention doc. number, if applicable)						
Equipment ID No.	**Critical/ noncritical**	**Name of equipment/ instrument**	**URS No. and approval date**	**DQ No. and approval date**	**IQ No. and approval date**	**OQ No. and approval date**	**PQ No. and approval date**	**Performance verification frequency**	**Last PV date**
AI-NNN	Critical	HPLC-1	Not available	AI/DQ/ NNN	AI/IQ/NNN	AI/OQ/ NNN	AI/PQ/ NNN	Yearly	dd/mm/yy
			–	dd/mm/yy	dd/mm/yy	dd/mm/yy	dd/mm/yy	dd/mm/yy	–
AI-NNN	Critical	GC-2	AI/URS/ NNN	Provided by vendor				Quarterly	
			dd/mm/yy	dd/mm/yy	dd/mm/yy	dd/mm/yy	dd/mm/yy	dd/mm/yy	dd/mm/yy
AI-NNN	Critical	Refrigerator	AI/URS/ NNN	Not applicable	Not applicable	Not applicable	AI/PQ/ NNN	Yearly	–
			dd/mm/yy	dd/mm/yy	dd/mm/yy	dd/mm/yy	dd/mm/yy	dd/mm/yy	dd/mm/yy
AI-NNN	Critical	pH meter	AI/URS/ NNN	AI/DQ/ NNN	AI/IQ/NNN	AI/OQ/ NNN	AI/PQ/ NNN	Yearly	–
			dd/mm/yy	dd/mm/yy	dd/mm/yy	dd/mm/yy	dd/mm/yy	dd/mm/yy	dd/mm/yy
AI-NNN	Critical	Conductivity meter	AI/URS/ NNN	AI/DQ/ NNN	AI/IQ/NNN	AI/OQ/ NNN	AI/PQ/ NNN	Yearly	–
			dd/mm/yy	dd/mm/yy	dd/mm/yy	dd/mm/yy	dd/mm/yy	dd/mm/yy	dd/mm/yy

(Continued)

Table 10.13 (Continued)

Equipment ID No.	Critical/ noncritical	Name of equipment/ instrument	Qualification status (mention doc. number, if applicable)					Performance verification frequency	Last PV date
			URS No. and approval date	DQ No. and approval date	IQ No. and approval date	OQ No. and approval date	PQ No. and approval date		
AI-NNN	Critical	Autoclave	AI/URS/ NNN	AI/DQ/ NNN	AI/IQ/NNN	AI/OQ/ NNN	AI/PQ/ NNN	Yearly	–
			dd/mm/yy	dd/mm/yy	dd/mm/yy	dd/mm/yy	dd/mm/yy	dd/mm/yy	dd/mm/yy
AI-NNN	Critical	Hot-air oven	AI/URS/ NNN	AI/DQ/ NNN	AI/IQ/NNN	AI/OQ/ NNN	AI/PQ/ NNN	Yearly	–
			dd/mm/yy	dd/mm/yy	dd/mm/yy	dd/mm/yy	dd/mm/yy	dd/mm/yy	dd/mm/yy
AI-NNN	Noncritical	Heating block	AI/URS/ NNN	Not applicable	Not applicable	Not applicable	AI/PQ/ NNN	Not applicable	–
			dd/mm/yy	Not applicable	Not applicable	Not applicable	Not applicable	Not applicable	–
AI-NNN	Critical	Karl Fischer titrators	AI/URS/ NNN	AI/DQ/ NNN	AI/IQ/NNN	AI/OQ/ NNN	AI/PQ/ NNN	Yearly	–
			dd/mm/yy	dd/mm/yy	dd/mm/yy	dd/mm/yy	dd/mm/yy	dd/mm/yy	dd/mm/yy
AI-NNN	Critical	Laminar flow hood	AI/URS/ NNN	AI/DQ/ NNN	AI/IQ/NNN	AI/OQ/ NNN	AI/PQ/ NNN	Six monthly	–
			dd/mm/yy	dd/mm/yy	dd/mm/yy	dd/mm/yy	dd/mm/yy	dd/mm/yy	dd/mm/yy

AI, analytical instrument; NNN, sequential number in three digits.

verification, calibration, and maintenance of equipment throughout its shelf life.

The main goal in qualifying the laboratory equipment is to ensure the validity of data. This calls for a robust instrument qualification program. Qualification shall involve science-based approach to provide documented evidence that the instrument is capable of consistently operating within established limits and tolerances.

Acknowledgments

This publication was supported/partially supported by Dr Cyrus (President, Lupin Ltd., Biotech Division, India), Dr V. Srini Srinivasan (Premier Expert Consultant, NDA Partners, LLC), Dr Behnam Davani (Principal Scientific Liaison, General Chapters, Science Division at US Pharmacopeia).

We are very grateful to Shalini Mandelia (QA Executive, Lupin Ltd., Biotech Division, India) for her contribution in writing the chapter and cited examples by reviewing the references.

We are also thankful to Archana Verma (QC Manager, Lupin Ltd., Biotech Division, India) for her contribution and for sharing her experience.

We are also immensely grateful to Ashok Khandivar (QA Senior Executive, Lupin Ltd., Biotech Division, India) for assisting in writing and significantly improving the overall write-up.

References

1 Qualification of Equipment, Annex 1: Qualification of HPLC Equipment, PA/PH/OMCL (11) 04 CORR.
2 Performance Verification Test of High Performance Liquid Chromatography, International Journal of Chromatographic Science Lantider Kassaye, Getachew Genete.
3 Dionex User Manual for Operational Qualification/Performance Qualification for HPLC Instruments.
4 User Manual, Waters Empower Software SQT Qualification Tool.
5 Primer on Analytical Instrument Qualification and System Validation, Agilent Technologies.
6 Performance of UV–Vis Spectrophotometers, Dr. Herman Lam, Glaxo Wellcome Inc. (Canada).
7 USP General Chapter <1058>: Analytical Instrument Qualification.
8 Qualification of Equipment, Annex 3: Qualification of UV–Visible Spectrophotometers, PA/PH/OMCL (07) 11 DEF CORR.
9 USP General Chapter <851>: Spectrophotometry and Light Scattering.

10 Qualification of Equipment, Annex 5: Qualification of Automatic Titrators, PA/PH/OMCL (07) 108 R6.
11 Qualification of Balances, Annex 8: Qualification of Equipment, PA/PH/OMCL (12) 77 7R.
12 Qualification of Equipment, Annex 6: Qualification of Piston Pipettes, PA/PH/OMCL (09) 64 2R.
13 ISO 8655: 2 Piston Operated Volumetric Apparatus.
14 Qualification of Equipment, Annex 2: Qualification of GC Equipment, PA/PH/OMCL (16) 17 R.
15 Qualification of Analytical Columns, PA/PH/OMCL (12) 128 5R.
16 User Manual, Melting Point Apparatus, EZ Melt: MPA 100 Stanford Systems.
17 USP General Chapter <741>: Melting Range or Temperature.

List of Abbreviations

AA	atomic absorption
AAMI	American Association for the Advancement of Medical Instrumentation
ACI	Andersen cascade impactor
ANADA	abbreviated new animal drug application
ANDA	abbreviated new drug application
ANSI	American National Standard Institute
APSD	aerodynamic particle size distribution
ASTM	American Society for Testing and Materials (now: ASTM International)
BAM	bacteriological analytical manual
BIO	Biotechnology Industry Organization
BLA	Biologics License Application
BSI	British Standards Institution
CAPA	corrective and preventive actions
CE	capillary electrophoresis
CFR	Code of Federal Regulations
cGMPs	current good manufacturing practices
CI	confidence interval
CQA	critical quality attributes
CS	computer system
CSA	Canadian Standards Association
DL	detection limit
DQ	design qualification
ELN	electronic lab notebook
EP	European Pharmacopeia)
EU	European Union
FCC	Food and Chemical Codex
FDA	Food and Drug Administration
FID	flame ionization detector
FPD	fine particle dose

Pharmaceutical Analysis for Small Molecules, First Edition. Edited by Behnam Davani.

FPM	fine particle mass
FPF	fine particle fraction
GC	gas chromatography
GDocP	good documentation practices
GLPs	good laboratory practices
GMPs	good manufacturing practices
HPLC	high-performance liquid chromatography
ICH	International Conference on Harmonization
ICH-Q	International Conference on Harmonization – Quality
ICP	inductively coupled plasma
ICP-AES	inductively coupled plasma–atomic emission spectrometry
ICP-MS	inductively coupled plasma–mass spectrometry
ICP-OES	inductively coupled plasma–optical emission spectrometry
IEC	ion-exchange chromatography
IPEC	International Pharmaceutical Excipients Council
IQ	installation qualification
IR	infrared
IRA	Interim Revision Announcement
IRB	Institutional Review Board
ISO	International Organization for Standardization
IVIVC	in vitro in vivo correlation
JP	Japanese Pharmacopeia (JP)
JP	Japanese Pharmacopoeia
k' or k	capacity factor or retention factor
LIMS	lab information management system
LOD	loss on drying
LOI	loss on ignition
LSL	lower specification limit
MMAD	mass median aerodynamic diameter
MPN	most probable number
MS	mass spectrometry
N	number of theoretical plates
NADA	new animal drug application
NARA	National Archives and Records Administration
NDA	new drug application
NF	national formulary
NMR	nuclear magnetic resonance
NMT	not more than
NPC	normal-phase chromatography
OOS	out of specification
OOT	out of trend
OQ	operational qualification

OTC	over the counter
P&IDs	piping and instrument diagrams
PAT	process analytical technology
PDA	Parenteral Drug Association
PDE	permissible daily exposure
PDG	Pharmacopeial Discussion Group
PF	Pharmacopeial Forum
PQ	performance qualification
PQIT	periodic quality indicator tests
PV	performance verification
QbD	quality by design
QC	quality control
QL	quantitation limit
QMS	quality management system
R	resolution
ROI	residue on ignition or sulfated ash
RPC	reversed-phase chromatography
SD or S	standard deviation
SEC	size-exclusion chromatography
SFC	supercritical fluid chromatography
SMF	site master file
SOP	standard operating procedures
T	tailing factor or asymmetry factor
t_0	retention time of the nonretained peak
TAMC	total aerobic microbial count
TLC	thin-layer chromatography
t_R	retention time of the analyte peak
TYMC	total yeast and mold count
URS	user requirement specification
US	United States
USAN	United States Adopted Names
USL	upper specification limit
USP	United States Pharmacopeia
UV	ultraviolet
Wb	peak width at the baseline

Index

Pharmaceutical Analysis for Small Molecules, First Edition. Edited by Behnam Davani.

r

s

www.ingramcontent.com/pod-product-compliance
Ingram Content Group UK Ltd.
Pitfield, Milton Keynes, MK11 3LW, UK
UKHW020214290926
15470UKWH00001B/186